PRICE DYNAMICS IN EQUILBRIUM MODELS

Advances in Computational Economics

VOLUME 16

Price Dynamics in Equilibrium Models

The Search for Equilibrium and the Emergence of Endogenous Fluctuations

by

Jan Tuinstra
University of Amsterdam

Springer Science+Business Media, LLC

Library of Congress Cataloging-in-Publication Data
Tuinstra, Jan.
Price dynamics in equilibrium models: the search for equilibrium and the emergence of endogenous fluctuations / by Jan Tuinstra.
p.cm. -- (Advances in computational economics ; v. 16)
Includes bibliographical references and index.
ISBN 978-0-7923-7265-3 ISBN 978-1-4615-1661-3 (eBook)
DOI 10.1007/978-1-4615-1661-3
1. Prices--Mathematical models. 2. Equilibrium (Economics)--Mathematical models. 3. Statics and dynamics (Social sciences) 4. Nonlinear theories. I. Title. II. Series.

HB221.T77 2000
338.5'2--dc21 00-051986

Originally published by Kluwer Academic Publishers in 2001

Printed on acid-free paper.

Contents

List of Figures

Preface

A long-standing unsolved problem in economic theory is how economic equilibria are attained. This book considers a number of adjustment processes in different economic models and investigates their dynamical behaviour. Two important themes arising in this context are "bounded rationality" and "nonlinear dynamics". The results presented here indicate that endogenous fluctuations are the rule rather than the exception in the search for equilibrium. The book is written for anyone with an interest in economic theory in general and bounded rationality and endogenous fluctuations in particular. It is entirely self-contained and accessible to readers with only a limited knowledge of economic theory.

Parts of the work presented here are based upon research projects with Edward Droste, Cars Hommes and Claus Weddepohl, whom I thank for allowing me to use the results of our joint work. Chapter 3 is based upon an article published by the *Journal of Economics-Zeitschrift für Nationalökonomie*. I thankfully acknowledge Springer-Verlag for permission of reproducing these results here. Furthermore, I benefitted greatly from vivid discussions with Maarten-Pieter Schinkel and Michel van de Velden and I am thankful to Cees Diks and Henk van de Velden for assistance with practical issues. I am grateful to Allard Winterink and his associates at Kluwer Academic Publishers for providing me with the opportunity to write this book.

Amsterdam, September 2000

Chapter 1

Introduction and Outline

Economic theory deals primarily with the existence of equilibria, i.e. prices at which market clearing occurs. If such equilibrium prices exist, the question arises how they are attained. The problem of how economic agents can coordinate on such an equilibrium has received much less attention than the problem of existence, but nevertheless is highly important. Indeed, if this problem cannot be solved satisfactorily, economic predictions and comparative statics based upon equilibrium analysis are strongly hypothetical. In this monograph we discuss, in a number of different equilibrium models, some different adjustment processes and try to gain some insights into out-of-equilibrium dynamics and the question whether coordination on an equilibrium will in general be achieved or whether other dynamical phenomena may be encountered.

The best-known adjustment process in general equilibrium theory is the tâtonnement process, which is a very simple way to model the *law of supply and demand.* This law of supply and demand states that the price of a commodity will increase when demand for that commodity exceeds supply and that the price will decrease if supply exceeds demand. It is by now well-known (see Arrow and Hahn (1971)) that convergence of the tâtonnement process to an equilibrium price vector can only be guaranteed under rather restrictive assumptions on the aggregate excess demand functions. Moreover, it has been shown that for reasonable specifications of the fundamentals of the economy the tâtonnement process might lead to endogenous periodic and chaotic fluctuations of prices around an unstable steady state.

In this monograph the tâtonnement process and a number of other adjustment processes are studied. These models differ significantly from each other but they have some important characteristics in common. Firstly, most of them are models of *bounded rationality*. The problem of coordination on equilibria becomes relatively easy once one assumes unbounded rationality of economic agents, which implies that they have perfect knowledge of their economic environment, including the behaviour of all other economic agents and with this information are able to compute the, possibly complicated, equilibria. This assumption of unbounded rationality is much too demanding however and hence we focus on models where economic agents use simple rules of thumb or models where they try to make some inference about their economic environment by looking at observations of certain economic variables. There is a "wilderness of bounded rationality" but we restrict this wilderness by considering rules of thumb that are "sensible" or "approximately rational".[1] Secondly, since price adjustment processes are in general highly nonlinear, we apply the methodology of *nonlinear dynamics*. In fact, we find that in most of our models the adjustment processes have typical nonlinear features, such as periodic and chaotic behaviour, just as can be encountered in the tâtonnement process. In the rest of this introductory chapter we will first briefly consider bounded rationality, then we will discuss the field of nonlinear economic dynamics and illustrate some of its features by considering a simple example of a tâtonnement process in a general equilibrium model. We will conclude with a brief outline of the remaining chapters in this monograph.

1.1 Bounded Rationality

Economics is rooted deeply in the theory of choice. Economic agents are decision makers choosing their actions, given some feasibility constraints, in a way that optimizes their objective function. The leading paradigm in economic theory states that economic agents are *rational* in the sense that they know the set of actions from which they can choose, they know the set of consequences associated to each of these actions and they have a complete preference ordering over these different consequences. The rationality assumption can be divided into two parts. Firstly, it is assumed that the agents exhibit individual rationality, that is, given the *perceived*

[1] In the words of Sargent (1993, p.3): "I interpret a proposal to build models with 'boundedly rational' agents .... by expelling rational agents from our model environments and replacing them with 'artificially intelligent' agents who behave like econometricians."

feasibility constraints economic agents choose that action that optimizes their objective function. Secondly, the perceived feasibility constraints have to be *mutually consistent* over all individuals. Since the feasibility constraints for any economic agent are in part determined by the actions of other economic agents this implies that agents have to predict the actions of all other agents. The rationality assumption thus also applies to expectations with respect to the actions of other agents and culminates in the so-called *rational expectations hypothesis.* This hypothesis is indeed very demanding and really presumes *unbounded* rationality, since it requires that each agent exactly knows how all other agents behave. It implies that all agents know the economic environment they are in and are able to compute equilibria and solve difficult optimization problems. It implicitly endows them with more ability and knowledge of the economic system, than economic theory can provide them with. Since the introduction of rational expectations by Muth (1961) and the introduction into macroeconomics by Lucas (1971) the rational expectations paradigm has been, until quite recently, the dominating paradigm in the modelling of expectations in economics.

The strong requirements on knowledge, information gathering and computational abilities implied by the rationality assumption recently have lead a number of scholars to study models of bounded rationality (see for example the review by Conlisk (1996) and the books by Sargent (1993, 1998)). Several reasons (apart from common sense) for this can be identified. There is a lot of experimental evidence indicating that people are not as rational as is often assumed by economic theorists (see for example Thaler (1992)). Furthermore, the existence of deliberation and information costs is often ignored. As argued above, rationality requires much computational effort and information gathering. Since these are costly activities, they should be part of the decision problem of economic agents and it might well be "rational" to decide not to invest too much time and effort in being able to make the best decision if the trade off between the increase in deliberation costs is not matched by an equal increase in pay-off (in, for example, profit or utility). Proponents of the rationality approach have defended their position by using the "as if" argument, which states that economic agents are not really rational, but act *as if* they are. This is supported by the argument that agents can learn how to behave rationally and that those agents who do not will be driven out of the market since they will perform not as well as the rational agents. Another argument for rational expectations is that it disciplines the way in which behaviour of agents is modelled as opposed to the "wilderness of bounded rationality" which emerges if the rationality assumption is relaxed. These arguments implicitly suggest that we have to study models of bounded rationality and analyze whether in such models agents eventually behave as if they were rational. These models might also discipline the way in which we model bounded rationality.

Models of bounded rationality assume that agents are not unboundedly rational. Some models of bounded rationality assume that agents use very simple rules of thumb in order to make decisions. An interesting issue then is which of these rules of thumb performs better than others and will be imitated by other agents and drives out rules of thumb that perform badly. This is the subject of evolutionary models, such as models from evolutionary game theory (see e.g. Samuelson (1997), Vega-Redondo (1996) and Weibull (1995)). Other models of bounded rationality assume that agents are individually rational but boundedly rational with respect to their perceptions of the economy. They make inferences about their economic environment on the basis of some (incomplete) information set, for example by estimating some, possibly mis-specified, model of the economy. On the basis of this perceived model of the economy they are individually rational in determining their optimal choices.

There is another reason why the models of bounded rationality have received attention in the economic literature, particularly in models with multiple rational expectations equilibria. In these models bounded rationality is used as an equilibrium selection device, to determine which of the different equilibria is most "likely". The models of bounded rationality studied in this monograph are meant to provide more insights in how agents behave in real life and should therefore not be considered as models of equilibrium selection. Models of bounded rationality are an interesting object of study because we believe economic agents in fact are boundedly rational.

1.2 Theory of Nonlinear Dynamics

In this monograph we analyze a number of dynamical systems, which in general will be nonlinear. As an approximation one could study the linearization of these dynamical systems around a steady state of the dynamics and consider local stability of this linearized system. However, this approach neglects important dynamical features of the original nonlinear system. In particular, the linearized system is only a valid approximation for the nonlinear system in the neighbourhood of the steady state. In contrast to the case of a linear system, in a nonlinear system a locally unstable steady state does not imply that the time paths will diverge since the nonlinearity of the dynamical system may cause (some or all of the) time paths to be bounded, which may lead to some kind of periodic or nonconverging but bounded behaviour.

In recent years the importance of the features of nonlinear dynamics has become gradually acknowledged in economic theory. In Section 1.2.1 we will briefly review the field of nonlinear economic dynamics and in Section 1.2.2 we will illustrate some important concepts from the theory of nonlinear dynamics, by studying a simple example of a tâtonnement process.

1.2.1 Nonlinear economic dynamics: a historical perspective

Nonlinear dynamic economic models can exhibit periodic cycles or even chaotic fluctuations. The question thus arises whether business cycles can be explained endogenously by nonlinearities in the underlying economic system. This would be an appealing alternative to the currently dominating paradigm that business cycles arise from exogenous shocks hitting an inherently stable (linear) system. In the last decades some progress has been made in this endogenous business cycle approach.

Among the earliest applications of the theory of nonlinear dynamics to economic models were the Cournot adjustment model by Rand (1978) and the growth models by Day (1982, 1983). These models have been criticized for the fact that agents in these models are behaving suboptimal and are in fact irrational. An important role in the development of nonlinear economic dynamics has subsequently been played by the overlapping generations model. Benhabib and Day (1982) and Grandmont (1985) have shown that in the overlapping generations model with utility maximizing agents, market clearing in every period and perfect foresight, periodic and chaotic behaviour can occur. These models have in common with the earlier contributions that they can be written as a one-dimensional difference equation. The existence of periodic and chaotic behaviour then requires that this difference equation is hump-shaped, which for the overlapping generations model corresponds to a strong conflict between income effects and intertemporal substitution effects. In recent years higher dimensional economic dynamic models have been studied and it appears that these models exhibit periodic and chaotic behaviour for more 'reasonable' specifications. Poincaré (1890) already pointed out that the existence of *homoclinic orbits* in such higher dimensional systems implies very complicated behaviour. De Vilder (1996), for example, has studied the overlapping generations model with capital and has shown the occurrence of chaotic behaviour in this model, for a specification of the model where income effects are dominated by substitution effects everywhere. Other recent examples of two dimensional economic models exhibiting strange behaviour are the cobweb model with heterogeneous expectations (Brock and Hommes (1997)), where evolutionary competition between different boundedly rational expectation rules leads to the existence of homoclinic orbits and the tâtonnement process with three commodities (Goeree, Hommes and Weddepohl (1997)) which for certain values of the speed of adjustment leads to chaotic fluctuations in prices.

The fact that periodic and chaotic behaviour can be consistent with utility maximization, market clearing and rational expectations or perfect foresight in a general equilibrium model can hardly be called surprising in light of the "anything goes" results by Sonnenschein (1973), Mantel (1974,1976) and Debreu (1974) who have shown that any set of continuous functions satisfying Walras' law and homogeneity of degree zero in prices, can be the excess demand functions for such an general equilibrium model with utility maximizing agents. One of the challenges for the field of nonlinear economic dynamics therefore lies in establishing that chaotic behaviour occurs for empirically relevant specifications of the fundamentals of the economy, such as utility functions and production technologies.[2]

The cycles and chaotic behaviour found in the overlapping generations model correspond to equilibria of a general equilibrium model and in that sense are not fundamentally different from the (stationary) equilibria economic theorists are used to studying. The recent upsurge of models of bounded rationality has lead to a renewed interest in the theory of nonlinear dynamics, from a different perspective. These models are populated by agents that behave adaptively: they make decisions on the basis of past observations. In general this leads to very complicated nonlinear dynamical systems, which often give rise to erratic disequilibrium behaviour. Most of the erratic dynamics in the models

[2]There is a large literature on detecting nonlinear structures in financial and economic time series (see for example Brock, Hsieh and LeBaron (1992)), which we will not touch upon in this thesis. The results to date are ambiguous. There has not been much evidence for low-dimension chaos in financial time series but the methods used are very sensitive to noise.

studied in this monograph are of this type. The main challenge for the theory of nonlinear economic dynamics lies in identifying the economic mechanisms that generate these erratic dynamics.

1.2.2 An example

We now briefly turn to a simple example of a tâtonnement process which features some of the dynamical phenomena that we will encounter throughout this monograph. Consider an economy with three commodities. The aggregate excess demand functions of the first two commodities are given by

$$\begin{aligned} z_1(p_1, p_2, p_3) &= \left(\frac{p_2}{p_1}\right)^{\alpha} + \left(\frac{p_3}{p_1}\right)^{2-\alpha} - 2 \\ z_2(p_1, p_2, p_3) &= \frac{p_1 + p_3}{p_2} - 2, \end{aligned}$$

where $\alpha \neq 0$. The aggregate excess demand for the third commodity then follows from Walras' Law. The unique equilibrium price vector is given by $(p_1^*, p_2^*, p_3^*) = (1, 1, 1)$. Now take the third commodity as a numeraire $(p_3 = 1)$ and consider the following tâtonnement process

$$\begin{aligned} p_{1,t+1} &= p_{1t} + \lambda z_1(p_{1t}, p_{2t}, 1) \\ p_{2,t+1} &= p_{2t} + \lambda z_2(p_{1t}, p_{2t}, 1), \end{aligned} \tag{1.1}$$

where $\lambda > 0$ is the speed of adjustment. Let us denote the tâtonnement process (1.1) by the mapping $F : \mathbb{R}_+^2 \to \mathbb{R}_+^2$. We are interested in the time path or orbit of an arbitrary initial price vector $\mathbf{p}_0$, under F, that is, the sequence of price vectors $\{F^t(\mathbf{p}_0)\}_{t=0}^{\infty}$ (where $F^t(\mathbf{p}_0) = F(F^{t-1}(\mathbf{p}_0))$ and $F^0(\mathbf{p}_0) \equiv \mathbf{p}_0$, for each $\mathbf{p}_0$) that is generated by the tâtonnement process. We are also interested in how the qualitative features of this orbit depend upon the parameters λ and α.

A first step in studying the dynamics of F is to consider a linearization of F around the equilibrium price vector $(p_1^*, p_2^*) = (1, 1)$. The linearized system is

$$\begin{pmatrix} \overline{p}_{1,t+1} \\ \overline{p}_{2,t+1} \end{pmatrix} = \begin{pmatrix} 1-2\lambda & \alpha\lambda \\ \lambda & 1-2\lambda \end{pmatrix} \begin{pmatrix} \overline{p}_{1t} \\ \overline{p}_{2t} \end{pmatrix},$$

where $\overline{p}_{it} \equiv p_{it} - p_i^*$. Let $\mu_1(\lambda, \alpha)$ and $\mu_2(\lambda, \alpha)$ be the eigenvalues of the linearized system (notice that they depend upon the parameters of the model). If both eigenvalues of this linearized system are smaller than 1

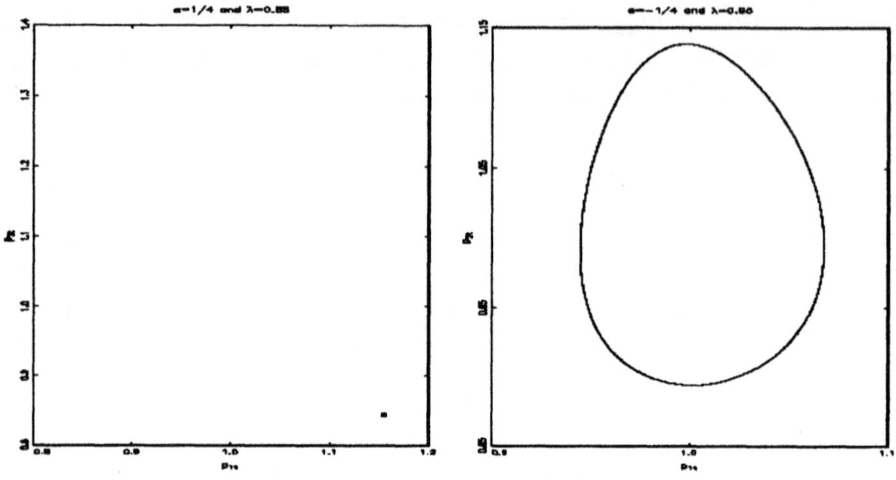

Figure 1.1: Long run behaviour for the tâtonnement process in the state space. On the horizontal axis p_{1t} and on the vertical axis p_{2t}. a) Period two orbit created in period doubling bifurcation. $\alpha = \frac{1}{4}$ and $\lambda = 0.85$. b) Quasi periodic cycle created in Hopf bifurcation. $\alpha = -\frac{1}{4}$ and $\lambda = 0.95$.

in absolute value then $(\overline{p}_{1t}, \overline{p}_{2t})$ will converge to $(0, 0)$. For the original system the steady state then is locally stable: for all price vectors in some small neighbourhood of the equilibrium price vector the adjustment process will converge to the equilibrium price vector. However, if at least one of the eigenvalues lies outside the unit circle the equilibrium is locally unstable. As one of the eigenvalues crosses the unit circle as a parameter varies, a so-called *bifurcation* occurs.[3] At such a bifurcation there is a *qualitative change in the dynamical behaviour* of the system. For example, the stability of the steady state changes and periodic cycles can be created. Figure 1.1 shows two important examples of what can happen at such a bifurcation. Figure 1.1.a) shows a period two cycle, for $\alpha = \frac{1}{4}$ and $\lambda = 0.85$ that is created in the *flip* or *period-doubling bifurcation* that occurs at $\alpha = \frac{1}{4}$ and $\lambda = \frac{4}{5}$. At these parameter values the eigenvalues are real and the smallest eigenvalue crosses the unit circle at -1 and a period two orbit of the dynamical system is created. The dynamical system then keeps on fluctuating between two price vectors.

[3]For a detailed analysis of the different local bifurcations, see for example the introductory texts on nonlinear dynamics by Guckenheimer and Holmes (1983) or Kuznetsov (1995).

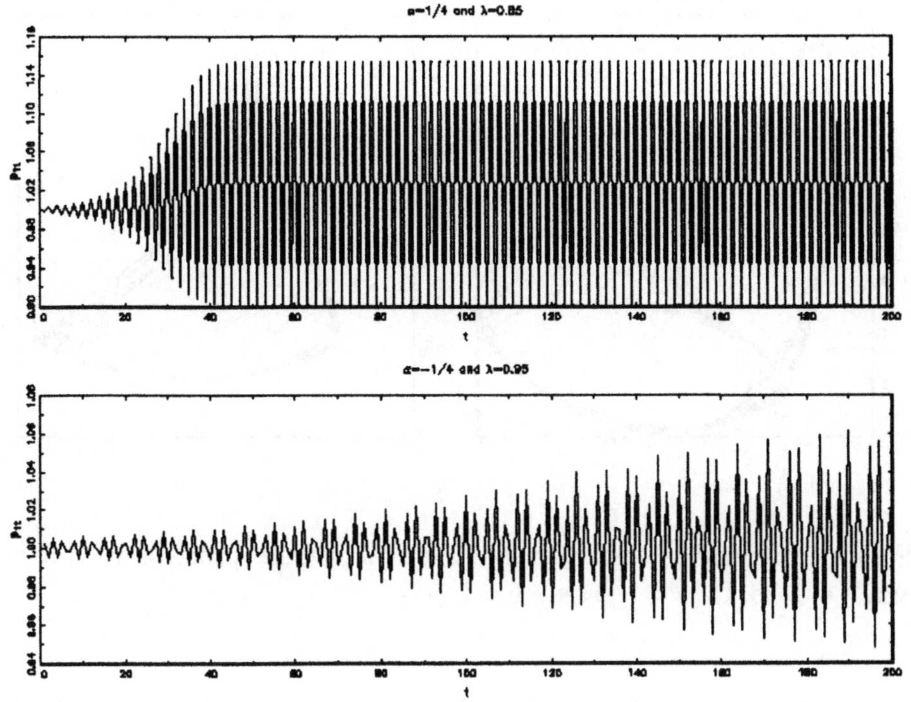

Figure 1.2: Time series of first 200 values of p_{1t} for a) $\alpha = \frac{1}{4}$ and $\lambda = 0.85$, and b) $\alpha = -\frac{1}{4}$ and $\lambda = 0.95$.

Figure 1.1.b) shows a closed curve which is invariant under F and which is created in the *Hopf bifurcation*[4] at $\alpha = -\frac{1}{4}$ and $\lambda = \frac{16}{17}$. At these values for the parameters, the eigenvalues are complex conjugates and are equal to 1 in absolute value.[5] The pictures in Figure 1.1 are constructed in the following way. An initial (nonequilibrium) price vector is chosen (in this case $p_{10} = 1$ and $p_{20} = 0.99$, which lies close to the equilibrium price vector) and F is applied to this initial price vector 6000 times and the last 5000 points (p_{1t}, p_{2t}) are plotted in the Figure. Figure 1.2 shows the corresponding time series for p_1. The pictures in Figure

[4] In fact, the Hopf bifurcation describes the creation of limit cycles in continuous time systems. For discrete time systems such as ours, it would be more appropriate to refer to this bifurcation as the *Neimark-Sacker bifurcation.*

[5] The bifurcation values can be easily computed as follows. The eigenvalues of the linearized system are $\mu_{1,2}(\lambda, \alpha) = 1 - 2\lambda \pm \lambda\sqrt{\alpha}$. They are real if $\alpha > 0$ and the smallest of them then is -1 for $\lambda = \frac{2}{2+\sqrt{\alpha}}$. They are complex conjugates whent $\alpha < 0$, and then they lie on the unit circle when $\lambda = \frac{4}{4-\alpha}$.

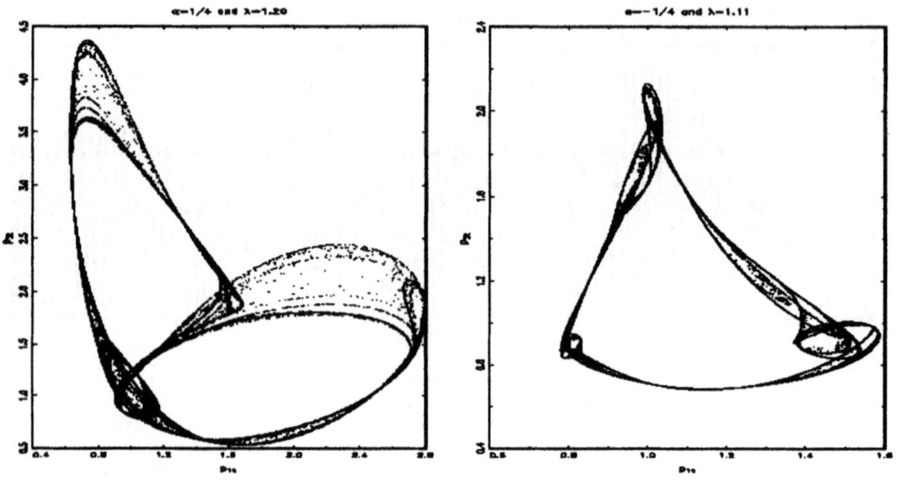

Figure 1.3: Strange attractors for the tâtonnement process. a) $\alpha = \frac{1}{4}$ and $\lambda = 1.20$, b) $\alpha = -\frac{1}{4}$ and $\lambda = 1.11$.

1.1 correspond to the long-run behaviour of the tâtonnement process (1.1), and from these pictures it is clear that for a nonlinear system perpetual endogenous fluctations around an equilibrium are a generic feature, after this equilibrium has become unstable. Such a set of points that captures the (possible) limit behaviour of the dynamical system is called an *attractor*.

Important features of an attractor are that it is an invariant set (that is, if some orbit is "trapped" in the attractor it will move over the attractor forever) and that it attracts all points in some neigbourhood. An attractor is in fact a generalization of the concept of a locally stable equilibrium. In the case of a stable equilibrium this attractor consists of one point. Other examples of attractors are a stable periodic orbit of period k, where the attractor consists of k different points or a stable closed curve, where orbits move over the closed curve quasiperiodically. Examples of these attractors are shown in Figure 1.1.

After the primary bifurcation of the equilibrium an infinite cascade of subsequent bifurcations can occur culminating in the existence of a *strange attractor*, with a complicated, fractal, structure. Examples are shown in Figure 1.3. From these pictures it is clear that the dynamics of the tâtonnement process for these parameter values behaves very erratically. These strange attractors can exhibit *sensitive dependence on*

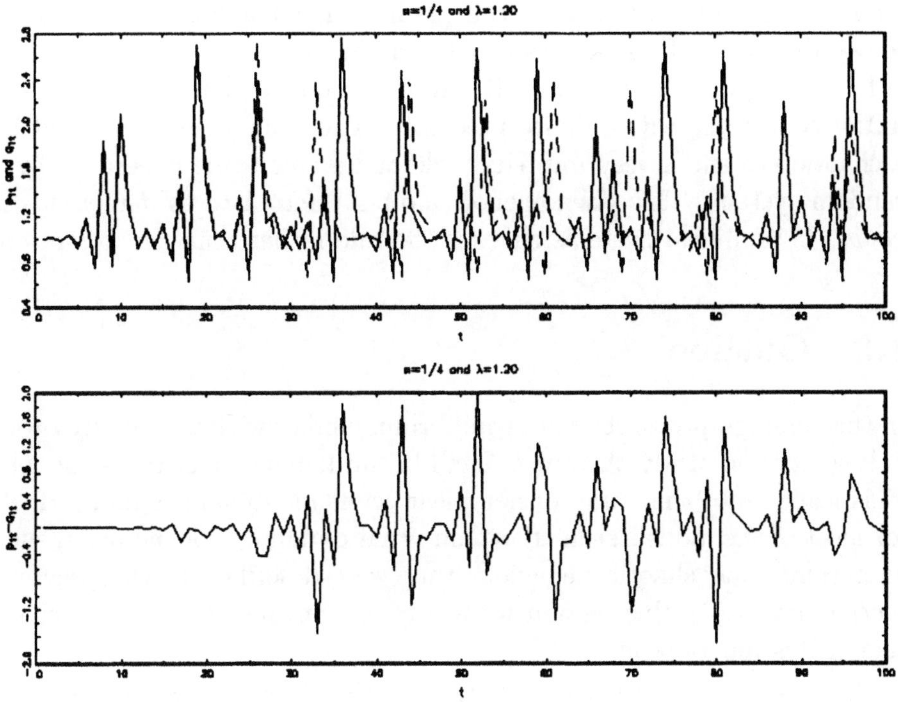

Figure 1.4: a) Two time series (for $\alpha = \frac{1}{4}$ and $\lambda = 1.20$) of first 200 values of p_{1t} for $(p_{10}, p_{20}) - (1, 0.99)$ (solid line) and for q_{1t} with $(q_{10}, q_{20}) = (1, 0.99001)$ (dotted line). b) Time series of difference $p_{1t} - q_{1t}$.

initial conditions. This concept can be clarified by considering time series corresponding to the strange and chaotic attractor in Figure 1.3.a). Two time series of orbits lying on this attractor and starting arbitrarily close to each other will follow quite different paths. To illustrate, consider Figure 1.4. In the first picture, two time series are shown. The time series with the solid line corresponds to the first 100 points iterated from the initial price vector $p_{10} = 1$ and $p_{20} = 0.99$, the time series with the broken line corresponds to the first 100 points iterated from an initial price vector $p_{10} = 1$ and $p_{20} = 0.99001$, which obviously starts out very close to the first one. However, it can be seen that the two orbits drift apart very fast. This is even more clear in the second picture which shows the time series of the difference between the two time series in the first picture. In the first 10 periods the time series seem to behave approximately the same, but after that their behaviour is radically different. This implies that long-run predictions for a chaotic system are impossible (even if one knows the deterministic system and is therefore able to compute the whole orbit) since it requires that one knows the initial conditions with infinite precision. The time series generated by a chaotic system is very unpredictable and seems to be generated by a random system. This phenomenon is often referred to as *deterministic chaos.* In nonlinear systems chaos is the rule rather than the exception.

1.3 Outline

In this monograph we consider equilibrium models with boundedly rational agents and study the out-of-equilibrium behaviour of these models. This leads us to the study of nonlinear economic dynamics and in fact, for most of the models studied, we find that dynamical phenomena, such as periodic and chaotic behaviour that were described in the previous section, occur. In this section we will briefly outline the different chapters of this monograph.

Chapter 2 deals with the analysis of the tâtonnement process in a simple general equilibrium model. This chapter builds on earlier work by Weddepohl (1996) and Goeree, Hommes and Weddepohl (1997), who have studied the tâtonnement process in exchange economies with two and three commodities, respectively, and have shown that strange behaviour might occur. Since the tâtonnement process requires that prices are normalized, they choose the price of one commodity (the numeraire) to be equal to 1, just as was done in the example from the previous section. However, which commodity is chosen, or in general, which price normalization rule is used, turns out to be important for the dynamics of the tâtonnement process and therefore the normalization rule has to be chosen with care. We focus on the simplex as a normalization rule. This normalization rule has as an advantage that, when it is combined with the discrete time multiplicative tâtonnement process, prices automatically stay on the simplex (much like the sphere is invariant for continuous linear tâtonnement process (see Arrow and Hahn (1971))). In this way no information is lost in choosing the normalization rule. We focus on a general equilibrium model with three commodities which implies that the tâtonnement process is a two-dimensional nonlinear dynamical system. Since there is a wide variety of possible economies, our approach is to focus on symmetric economies first. It appears that different kinds of dynamic behaviour correspond to different kinds of symmetry. In particular, we can have *symmetry breaking bifurcations* and *coexistence of attractors.* This coexistence of attractors also appears for closeby asymmetric tâtonnement processes.

In Chapters 3 and 4 we study the overlapping generations model. As was discussed in the previous section, this model has played an important role in the development of the theory of nonlinear economic dynamics. The papers by Benhabib and Day (1982) and Grandmont (1985) show that, under perfect foresight and market clearing, deterministic cycles and even chaotic behaviour can occur in the overlapping generations model. Chapter 3 deals with the close connection that exists between the appearance of such cycles and the existence of asymmetric equilibria in symmetric general equilibrium models of the kind that are analyzed in Chapter 2. These cycles are equilibrium cycles, that is, demand and supply along these cycles are in equilibrium. We construct some examples for which we can show that cycles of any period exist. Again it is of interest how agents can coordinate on such an equilibrium cycle (or on an equilibrium steady state). Chapter 4 deals with this problem of coordination and learning in the overlapping generations model. A simple overlapping generations economy with two generations and one commodity and boundedly rational agents is analyzed. The agents try to predict inflation rates by running a least squares regression on past prices or past inflation rates. The outcome of this learning process very much depends upon the way in which agents learn. If they learn on the basis of inflation rates beliefs may converge to some kind of limit belief, whereas the corresponding nonlinear dynamical system that determines the inflation rate still gives rise to periodic or chaotic behaviour. We call such a situation a *Beliefs Equilibrium,* since, given the class of beliefs they consider, the agents have no incentive to change their beliefs, although inflation rates might keep fluctuating in such a beliefs equilibrium.

In Chapters 2, 3 and 4 the focus has been on simple general equilibrium models with perfect competition. In these models it is unclear how prices are determined since all agents take prices as given. In Chapters 5 and 6 we study models of imperfect competition. Although we subscribe to the view that a real understanding of economic reality requires a general equilibrium perspective, the theory of general monopolistic equilibrium is quite complicated and includes a number of pitfalls. Hence, we will study the imperfect competition models mainly in a partial equilibrium context. Chapter 5 deals with what is probably the best-known partial equilibrium model of imperfect competition: the homogenous Cournot duopoly model. The question is whether agents can learn to play the Cournot-Nash equilibrium by using simple behavioural rules. We address this issue by developing a model of Cournot competition where there is evolutionary competition between these different rules. That is, there is a population of firms playing the Cournot duopoly game in pairs and the fraction of agents using a certain rule increases when average payoff corresponding to that rule is high in comparison with payoffs for other rules. We find that if agents switch relatively easy between different rules this evolutionary competition can generate complicated behaviour. In fact, for one typical example we are able to establish the occurrence of a *homoclinic bifurcation* between the stable and unstable manifolds of the equilibrium saddle point. This implies the existence of strange attractors for this model. Hence, evolutionary competition between a costly sophisticated behavioural rule and a cheap habitual rule of thumb leads to chaotic fluctuations. This kind of behaviour seems to be robust against considering different approaches to modelling evolutionary competition.

Chapter 6 reviews several models where firms set prices instead of quantities. The general setup is as follows. There are a number of firms producing a differentiated commodity. There is incomplete information in the following sense: firms do not have exact knowledge of the demand curve they are facing. Each firm acts as if it is a monopolist for the commodity it produces and neglects the effects from prices set by the other firms. Each firm then tries to estimate its demand curve and on the basis of this perceived demand curve sets a price for his product. This gives the firm new information, leading to a new estimate of the demand curve and subsequently to a new price. Notice that beliefs are misspecified, since firms do not take other firms prices into account. There might also be a misspecification in the structural form of the estimated relationship. We consider different learning models. One model assumes that firms have a lot of local information on the demand curve, it knows the amount it could have sold against last periods price and the price elasticity of the demand curve at that price. It uses this information (from the previous period) to estimate the demand curve. It can then be shown that this adjustment process converges to the Bertrand-Nash equilibrium if the cross-price effects are relatively small. If these become too large, periodic and complicated behaviour can arise. Such a learning process only uses the most recent observations. We also discuss the model by Kirman (1983) who considers a model where all previous observations on prices and quantities combinations (but not on price elasticities) are used to estimate the individual demand function. Prices and quantities then seem to converge to a so-called conjectural equilibrium. There is a continuum of these conjectural equilibria, and therefore prices and quantities in general do not converge to the Bertrand-Nash equilibrium.

In this monograph we consider a number of models from economic dynamics. These models are in general nonlinear and most of them assume that agents are boundedly rational. In general, the economic environment is too complex to be perfectly known. Agents therefore will often turn to simple habitual rules of thumb or adaptively updated misspecified models of the world to choose the appropriate actions. Through trial-and-error learning or evolutionary competition between the different rules of thumb boundedly rational agents may still end up in some kind of equilibrium, such as the Beliefs-Equilibria discussed in Chapter 4, which in some sense is "approximately rational". On the other hand, these adaptive models open the door for complicated dynamics. In fact, chaotic dynamics seem to be the rule rather than the exception.

Chapter 2

A Tâtonnement Process

2.1 Introduction

One of the simplest and best-known price adjustment processes is the so-called tâtonnement process. In this chapter we will study such a process in a general equilibrium model with three commodities, which gives rise to a two-dimensional nonlinear dynamical system. Partly due to the growing attention for the study of nonlinear economic dynamics there has been a renewed interest in the discrete tâtonnement process in recent years. It has been shown that the tâtonnement process does not always converge to an equilibrium price vector and can exhibit periodic or chaotic behaviour (Bala and Majumdar (1992), Day and Pianigiani (1991), Goeree, Hommes and Weddepohl (1997), Saari (1985) and Weddepohl (1995)). Our approach differs from previous contributions in that we choose the simplex as a price normalization rule instead of taking some good as numeraire. We argue that the simplex is a natural price normalization rule since it is an invariant space of the multiplicative adjustment process, just as the unit sphere is an invariant space of the continuous time linear adjustment process.

We are interested in the dynamics of the tâtonnement process for all possible economies. However, general equilibrium models have too many free parameters and therefore as a first step towards a systematic analysis, we focus on tâtonnement processes for symmetric general equilibrium models. We apply the literature on dynamical systems with symmetries (see Golubitsky, Stewart and Schaeffer (1988)) to study the resulting symmetric tâtonnement processes.

Since some economic models have a natural symmetry, our study of symmetric dynamical systems also has some importance from a methodological viewpoint. Examples of symmetries arising naturally in economic models are the cyclical symmetry in overlapping generations models (this will be the subject of the next chapter), the reflectional symmetry in the cobweb model with heterogeneous beliefs and linear supply and demand curves (see Brock and Hommes (1997)) and the reflectional symmetry in some specifications of the evolutionary model of Cournot competition as discussed in Chapter 5 of this monograph.

After reviewing the existing literature on price adjustment processes briefly in Section 2.2 we present our model and show results from numerical simulations in Section 2.3. In Section 2.4 we introduce symmetry. In Section 2.5 we look at the generic local bifurcations of the equilibrium price vector given a certain symmetry. Among other things we see that so-called symmetry breaking bifurcations can occur. In Section 2.6 the global dynamics of symmetric tâtonnement processes are studied. In Section 2.7 an asymmetric tâtonnement process which is in some sense close to a symmetric process is analyzed to illustrate the relevance of studying symmetric processes for processes that have no nontrivial symmetries. In Section 2.8 we briefly discuss the multiplicity of equilibria and Section 2.9 concludes.

2.2 Price adjustment models

We want to study a price adjustment process in a competitive exchange economy. Let there be m consumers and n commodities. Each consumer takes prices as given and solves the optimization problem

$$\max_{x_{i1},...,x_{in}} U_i\left(x_{i1},...,x_{in}\right), \qquad i = 1,...,m, \tag{2.1}$$

subject to his or her budget constraint

$$\sum_{j=1}^{n} p_j x_{ij} \leq \sum_{j=1}^{n} p_j w_{ij}, \qquad i = 1,...,m, \tag{2.2}$$

where $U_i\left(.\right)$ is the strictly quasi-concave, strictly monotonic utility function of consumer i, x_{ij} the amount of good j consumed by consumer i, p_j the price of good j and w_{ij} the initial endowment of good j owned by consumer i. Let $x_{ij}\left(\mathbf{p}, \mathbf{w}_i\right)$ be the resulting demand for good j

by consumer i, where $\mathbf{p} = (p_1, ..., p_n) \in I\!R^n_+$ is the price vector and $\mathbf{w}_i = (w_{i1}, ..., w_{in}) \in I\!R^n_+$ the vector of initial endowments of consumer i.

Aggregate excess demand for good j is

$$z_j(\mathbf{p}) = \sum_{i=1}^{m} x_{ij}(\mathbf{p}, \mathbf{w}_i) - \sum_{i=1}^{m} w_{ij}, \qquad j = 1, ..., n. \qquad (2.3)$$

The excess demand functions satisfy Walras' Law, $\sum_{j=1}^{n} p_j z_j(\mathbf{p}) = 0$ or in vector notation: $\mathbf{p} \cdot \mathbf{z}(\mathbf{p}) = 0$, where a dot denotes the inner product and $\mathbf{z}(\mathbf{p})$ is the vector of aggregate excess demand functions. Furthermore the excess demand functions are homogeneous of degree zero in prices, i.e. $z_j(\alpha \mathbf{p}) = z_j(\mathbf{p}), \forall \alpha > 0$. Differentiating this last equality with respect to α at $\alpha = 1$ gives Euler's equation: $\sum_{k=1}^{n} p_k \frac{\partial z_j(\mathbf{p})}{\partial p_k} = 0$. The economy is in equilibrium at the equilibrium price vector $\mathbf{p}^*$ if $z_j(\mathbf{p}^*) = 0$ for all j. Sufficient conditions for such a general equilibrium to exist are the strict quasi-concavity and strict monotonicity of the utility functions and the requirement that aggregate initial endowments lie in the interior of the commodity space. It can furthermore be shown that generically the equilibria are locally unique and finite in number (Debreu (1970)) and that generically their number is odd (Dierker (1972)).

If at certain prices the economy is not in equilibrium there must be at least one agent who does not consume his optimal consumption level. It is a central assumption in economic theory (for example expressed by the notion of the *law of supply and demand*) that there are economic forces that drive the economy towards equilibrium. The well-known tâtonnement process is the simplest way to model these equilibrating forces and can serve as a first approximation to more realistic adjustment processes. This process can be summarized by the following two characteristics.

- Trade occurs if and only if all markets are in equilibrium.
- If the economy is not in equilibrium at a certain price vector, prices are adjusted according to the following rule. If there is excess demand for a good its price goes up, if there is excess supply for a good its price goes down and if the market for a good is in equilibrium its price remains the same.

The tâtonnement process can be formalized as a discrete or as a continuous dynamical system

$$p_{j,t+1} - p_{j,t} = F_j\left(p_{j,t}, z_j\left(\mathbf{p}_t\right)\right), \qquad j = 1, ..., n, \tag{2.4}$$

$$\frac{dp_j(t)}{dt} = F_j\left(p_j(t), z_j\left(\mathbf{p}(t)\right)\right), \qquad j = 1, ..., n, \tag{2.5}$$

with $F_j(p_j, z_j)$ a sign-preserving function in z_j that satisfies $F_j(p_{j,t}, 0) = 0$.

In order to discuss the stability properties of the tâtonnement process we distinguish local and global stability. An equilibrium price vector $\mathbf{p}^*$ is said to be a locally stable equilibrium of a certain tâtonnement process if the trajectories of all initial price vectors in some small neighbourhood of $\mathbf{p}^*$ converge to $\mathbf{p}^*$. $\mathbf{p}^*$ is called globally stable if the trajectories of all initial price vectors converge to $\mathbf{p}^*$. An economy with multiple equilibria obviously cannot have a globally stable equilibrium, but the tâtonnement process then is called *system stable* if the trajectories of all initial price vectors converge to *some* equilibrium price vector.

The continuous tâtonnement process has been extensively studied in the literature. The classical reference is Arrow, Block and Hurwicz (1959). Reviews can be found in Arrow and Hahn (1971), Hahn (1982) and Takayama (1985). Arrow, Block and Hurwicz (1959) consider the continuous tâtonnement process $F_j(p_{j,t}, z_j(\mathbf{p}_t)) = \lambda_j z_j(\mathbf{p}_t)$ and try to prove global stability of the equilibrium by constructing an appropriate Lyapunov function. They find that a sufficient condition for global stability is that all goods are gross substitutes, that is: $\frac{\partial z_j}{\partial p_k}(\mathbf{p}) > 0$, $\forall j$, $\forall k \neq j$, $\forall \mathbf{p}$.[1] This condition also implies that the equilibrium price vector is unique. An alternative condition is the weak axiom of revealed preference for aggregate demand functions. This weak axiom says that for any pair of price vectors $\mathbf{p}$ and $\mathbf{q}$ we must have: if $\mathbf{p} \cdot \mathbf{z}(\mathbf{q}) \leq 0$, then $\mathbf{q} \cdot \mathbf{z}(\mathbf{p}) > 0$. Both conditions can be shown to imply that $\mathbf{p}^* \cdot \mathbf{z}(\mathbf{p}) > 0$ for any $\mathbf{p} \neq \mathbf{p}^*$, that is, the value of the aggregate excess demands weighted with the equilibrium prices is always positive outside of equilibrium. The

[1] For small values of n this assumption can be relaxed: for $n = 2$ the tâtonnement process is always system stable and for $n = 3$ a sufficient condition for system stability of the tâtonnement process $\frac{d\mathbf{p}}{dt} = \mathbf{z}(\mathbf{p})$ is the so-called weak law of market demand, which states that $\sum_{i=1}^{n} \frac{\partial z_i}{\partial p_i}(\mathbf{p}) < 0$ for all $\mathbf{p}$ (Keenan and Rader (1985)).

Euclidian distance between the current price and the equilibrium price vector, that is $V(\mathbf{p}) = \sum_{j=1}^{n} \left(p_j - p_j^*\right)^2$, can then be taken as a Lyapunov function. We have

$$\begin{aligned}\frac{dV}{dt} &= 2\sum_{j=1}^{n} \left(p_j - p_j^*\right) \frac{dp_j}{dt} = 2\sum_{j=1}^{n} p_j z_j(\mathbf{p}) - 2\sum_{j=1}^{n} p_j^* z_j(\mathbf{p}) \\ &= -2\sum_{j=1}^{n} p_j^* z_j(\mathbf{p}) < 0,\end{aligned}$$

where the second equality follows from the specification of the adjustment process and the third from Walras' Law.[2] Since the time derivative is negative everywhere outside of equilibrium, prices converge to the unique equilibrium price vector. Arrow, Block and Hurwicz also consider a Lyapunov function corresponding to the "max" norm, for which more general results can be achieved. In particular, global stability can be shown for general sign preserving functions of aggregate excess demand and it can be shown that a sufficient condition for global stability is that the Jacobian matrix has a dominant diagonal, that is $\left|\frac{\partial z_j}{\partial p_j}(\mathbf{p})\right| > \sum_{k \neq j} \left|\frac{\partial z_j}{\partial p_k}(\mathbf{p})\right|$, for all j and all $\mathbf{p}$.

The discrete tâtonnement process has received much less attention than the continuous process. The first study of the stability of a discrete tâtonnement process was done by Uzawa (1959). He looked at the price adjustment process $p_{j,t+1} = \max\{p_{j,t} + \lambda z_j(\mathbf{p}_t), 0\}$ and proved that for *sufficiently small* $\lambda > 0$ gross substitutability is a sufficient condition for global stability. Therefore for the discrete tâtonnement process the stability of competitive equilibrium is determined by the properties of the aggregate excess demand functions as well as the specific form of the adjustment rule. Recall that the specification of the adjustment rule has not been derived from economic fundamentals but has to be postulated in an ad-hoc fashion.

The assumption of gross substitutability (or its alternatives) is rather severe. There are some well-known examples (e.g. Scarf (1960)) for which this condition does not hold and for which the continuous tâtonnement process does not converge to an equilibrium price vector, but where prices move over a cycle. Moreover, the possibility of nonconvergence of the tâtonnement process should not be surprising in the

---

[2] Without loss of generality we have taken $\lambda_j = 1$, for all j. This can be achieved by choosing the unit of each good appropriately.

light of the so-called Sonnenschein-Mantel-Debreu results (see Debreu (1974), Mantel (1974,1976) and Sonnenschein (1973)). These authors have shown that any set of continuous functions defined for relative prices bounded away from zero and satisfying Walras' Law and homogeneity can be the aggregate excess demand functions of an economy with utility maximizing agents with convex preferences, provided that the number of agents is at least as large as the number of commodities. Therefore no restrictions other than continuity, Walras' Law and homogeneity in prices can be imposed on the aggregate demand functions.

In search for an adjustment process that converges for almost all aggregate excess demand functions Smale (1976) suggests the following process

$$J(\mathbf{p})\frac{d\mathbf{p}}{dt} = -\lambda(\mathbf{p})\,\mathbf{z}(\mathbf{p}),$$

where $J(\mathbf{p})$ is the Jacobian matrix of the aggregate demand functions evaluated at price vector $\mathbf{p}$ and where the sign of $\lambda(\mathbf{p})$ depends on the sign of det $J(\mathbf{p})$. Smale shows that, under some boundary and regularity conditions it converges for almost all initial conditions. Notice that it implies that the auctioneer needs to have a lot more information on aggregate excess demand functions in order to update prices. In fact, in order to adjust the price for one particular commodity information on the excess demands for all other commodities is required. This goes against the intuition of the law of supply and demand, where price adjustments only depend on the own excess demand. Therefore the adjustment process suggested by Smale should be seen mainly as an algorithm to find an equilibrium and not as a realistic model of price adjustment.

This has led Saari and Simon (1978) to investigate what the minimal informational requirements are for an *effective price mechanism,* that is, a price adjustment process that converges for almost all possible economies. They consider price mechanisms of the following form (after a suitable normalization of prices)

$$\frac{d\mathbf{p}}{dt} = M\left(\mathbf{z}(\mathbf{p}), \frac{\partial z_1}{\partial p_1}, \ldots, \frac{\partial z_1}{\partial p_{n-1}}, \ldots, \frac{\partial z_{n-1}}{\partial p_{n-1}}\right),$$

and then ask the question of how many of the demand derivatives $\partial z_i/\partial p_j$ are necessary in order for $M(.)$ to be an effective price mechanism. They show that for $n \geq 3$ most of these derivatives are necessary. Hence the information requirements for an effective price mechanism are substantial.

From this it follows that one would only expect tâtonnement-like processes to converge for a small set of possible economies. In fact, this leads to the conjecture that all kinds of dynamical phenomena are possible in the tâtonnement process and indeed in recent years there have been a number of contributions that show the existence of complicated phenomena in the tâtonnement process, such as periodic, quasi periodic and chaotic behaviour (for example Bala and Majumdar (1992), Day and Pianigiani (1991), Goeree, Hommes and Weddepohl (1997), Saari (1985) and Weddepohl (1995)). All these papers consider discrete versions of the tâtonnement process but from our discussion above it should be clear that the same phenomena arise in continuous versions of the model. In this chapter we will also study a discrete tâtonnement process but our main focus will not be on the fact that strange behaviour is possible, but on a characterization of this behaviour. But first we discuss some other problems related to tâtonnement processes.

Besides the fact that it has poor stability qualities there are some conceptual problems with the tâtonnement process. The first problem is that there is no trade at disequilibrium prices. Clearly this is an unpalatable assumption and a realistic model of price adjustment would incorporate the possibility of disequilibrium trade. In particular, non-convergence of the tâtonnement process, as we will encounter in the present chapter, implies that trade never takes place. An alternative interpretation of the absence of trade at disequilibrium prices would be the following. There is a sequence of periods. Each period starts with the same allocation of initial endowments. All commodities are perishable and there are no means of saving. Then in each period a price vector emerges according to the tâtonnement process described above. In general this price vector will not coincide with an equilibrium price vector. Now (disequilibrium) trade takes place and the traded commodities are consumed. The next period then starts with the same allocation of initial endowments but with the (disequilibrium) price vector that was established in the current period.

Some alternative adjustment processes, so-called non-tâtonnement adjustment processes, have been suggested in which trade takes place at "false prices". We will now briefly discuss two of these. First consider the *Edgeworth process* (see Hahn (1962) and Uzawa (1962)). According to this Edgeworth process, in disequilibrium agents engage in trade with each other, that is feasible (that is, for each agent his trade has to satisfy his budget constraint) and that increases their utility. Endowments change and prices adjust according to excess demand. Then a next round of trading emerges, and so on. By taking the sum of all utilities as a Lyapunov function it can then be shown that this process converges to a Pareto efficient allocation, since the utilities of agents increase as long as trade takes place and per definition trade stops at such a Pareto efficient allocation. There are some difficulties with this process. First of all it is assumed that agents can always find people to trade with. This might become rather complicated, since situations might arise requiring a large group of consumers to interact, in order to perform mutually advantageous trade. Another problem is that speculative trade is assumed away. One could imagine a case where agents buy certain commodities only because they belief the price of this commodity will increase in the future and therefore will make them better of. An alternative non-tâtonnement process is the so-called *Hahn process* (Hahn and Negishi (1962), for a discussion and extensions see Fisher (1983)). In contrast to the Edgeworth process it is assumed in the Hahn process that there are well organized markets for all traded commodities. Trade then takes place in a way such that no agent has an individual excess demand (supply) for a commodity when there is aggregate excess supply (demand) for that commodity. As a Lyapunov function the sum of the *target* utilities is used. The target utility is the utility an agent obtains if he can execute all his desired trades. Prices react to excess demand in the normal way and therefore a commodity for which the agent has an individual excess demand (and hence wants to buy) becomes more expensive (since there is also an aggregate excess demand) and a commodity that the agent has an individual excess supply of and hence wants to sell becomes cheaper (since there is also aggregate excess supply). Therefore the agents always is worse of after a price change and its target utility decreases until it reaches a lower bound.

For both the Edgeworth and the Hahn process there is convergence under a mild set of conditions. The problem with both processes is that they converge to an allocation which is Pareto efficient but which, in general, does not correspond to a Walrasian equilibrium of the original economy. This is due to the fact that the incorporation of trade into the model continually changes the endowments of the consumers and hence the Walrasian equilibria corresponding to these endowments change. These processes are therefore not appropriate to study the stability of the original Walrasian equilibrium.

Another problem with the tâtonnement process arises when we ask the question who adjusts the prices. In the perfectly competitive economy described above all agents take prices as given in determining their optimal consumption. If the economy is not in equilibrium there are profit opportunities for economic agents since they can benefit from charging a higher price for goods in excess demand. Therefore it can be argued that perfect competition is only a good description of economic behaviour if the economy is in equilibrium (see Arrow (1959)). The traditional solution to this problem is the introduction of an *auctioneer* who sets prices. The tâtonnement process now works as follows: the auctioneer states an arbitrary price vector and collects information on the aggregate excess demands given this price vector. He then adjusts the price vector according to the rule stated above. This process is repeated until an equilibrium price vector is found. However, where all decisions of the consumers arise from utility or profit maximization the behaviour of the auctioneer is postulated ad hoc. This seems to be inconsistent with the general equilibrium framework. Whose behaviour does the auctioneer represent? It might be argued that a realistic model of price adjustment requires that prices are set by economic agents, in such a way that they believe it maximizes their utility or profit. In order to study these kinds of models we have to consider general equilibrium models with monopolistic competition. Here we will briefly review this literature.[3] There have been two main branches of this literature. One uses the subjective demand approach, following the seminal paper by Negishi (1961) and the other uses the objective demand approach, following the paper by Gabszewicz and Vial (1972).

Negishi (1961) considers an economy with two types of producers: perfect competitors, who take all prices as given and imperfect competitors who can affect some of the prices. It is assumed that for each commodity there is at most one producer able to affect its price. This commodity may be an input as well as an output for this producer. Each imperfect competitor has some conjecture over the demand curves of the commodities for which it can influence the price. These conjectures are assumed to be linear functions of the own price. The only restriction on the conjectures is that they have to pass through the actual price-quantity combinations that correspond to the current state of the economy. The economy then is in equilibrium if all consumers and perfectly competitive firms maximize their utility and profits given their budget restrictions and production sets, if all imperfectly competitive firms are maximizing their profits given their conjectures and production sets and if excess demand is zero. It can then be shown that, given standard as-

[3]For comprehensive surveys see Hart (1985) or Bonanno (1988).

sumptions on preferences and technology, such an equilibrium exists. A problem with the subjective demand approach is that it has no predictive power: if conjectures of imperfectly competitive firms are not restricted any further almost any allocation can be an equilibrium. For example, if conjectures are such that firms believe demand to be infinitely elastic at a certain price, the Walrasian equilibrium results. For this reason a number of people have tried to construct general equilibrium models of monopolistic competition that incorporate *objective* demand curves. These objective demand curves exactly tell the firms the demand they are facing.

In order to construct these objective demand functions we have to deal with the following problem. Firms want to choose prices or quantities in order to maximize profits. These profits are then redistributed to the agents who own the firms. These redistributed profits enter the budget equations of the agents and therefore are an ingredient of the objective demand functions that are used in the maximization process of the firms to generate these same profits. This circularity of profits is called the *feedback effect* and it clearly complicates the optimization problem of the firms. A related problem is whether profit maximization is the right criterion for oligopolistic firms. Since the owners of the firm can influence prices there might be some trade-off between lower nominal profits and more favourable prices. So maximization of utility of the owners might be a better criterion than profit maximization. However, aggregation problems might occur if different owners have different preferences.

In the subjective Negishi framework it does not matter whether firms are price setters or quantity setters, since they act as monopolists and therefore arrive at the same decision in both cases. In the objective demand approach it does matter, since firms take account of the actions of the other firms in the economy. Gabsewicz and Vial (1972) construct a general oligopolistic equilibrium model in the Cournot-Nash framework in the following way. All firms are imperfect competitors producing commodities using some "nonmarketable" resources of the consumers. There are no intermediate goods, as in the Negishi model. The commodities produced by the firms are distributed to the consumers (so it is not profits but commodities themselves that are distributed). The objective inverse demand functions are then constructed in the following way. For given production plans of the firms the consumers have modified initial endowments, consisting of the original initial endowments and the redistributed produced commodities. These consumers then form an exchange economy for which an equilibrium can be computed, which of course depends upon the production plans of the firms. The equilibrium prices of this exchange economy as a function of these production plans constitute the objective inverse demand functions. The general oligopolistic equilibrium (called a Cournot-Walras equilibrium by Gabsewicz and Vial) then corresponds to the Cournot-Nash equilibrium of the game played by the firms and the equilibrium of the corresponding exchange economy. There are several problems with this approach. In particular, the equilibrium of the exchange economy does not have to be unique for a given production plan and the objective inverse demand function is not necessarily continuous.

Nikaido (1975) constructs an objective demand model with price setting behaviour. The production sector is represented by a Leontief technology, which is rather restrictive but allows for the introduction of intermediate goods. Consumers have only an endowment of labour and sell this to the firms to generate income in order to consume other goods. All profits go to the owners of the firms, the capitalists, who use this profit to consume goods. Total demand then consist of final demand by workers, final demand by capitalists and intermediate demand by firms. In equilibrium this equals the production of the firms. For any vector of prices this equilibrium condition then gives a production vector which is consistent with it. In this way objective demand functions can be constructed and an equilibrium then corresponds to a Bertrand-Nash equilibrium of the corresponding game that is played between the firms. Notice that there is one firm per sector in this model, or alternatively, each sector behaves as if it was governed by one firm.

The objective demand approach also has some drawbacks. In particular, it seems to be rather farfetched to assume that all firms construct a complete general equilibrium model in order to determine the objective demand curves. A compromise between the objective and subjective approaches would be to consider a subjective demand approach that incorporates more consistency conditions than Negishi (1961) does. An example is provided by Silvestre (1977). He considers a version of the Nikaido model and requires that in equilibrium not only subjective demand has to equal objective demand, but, at this equilibrium, the slope of the subjective demand curve also has to coincide with the slope of the objective demand curve. This can be extended to higher-order derivatives (Gary-Bobo (1987)). Other examples of relaxing the informational content of objective demand are given by Hart (1985) and Marschak and Selten (1974). In their models the firms treat the distribution of wealth as given in order to deal with the circularity of profits. Two other problems in the objective demand approach are that an equilibrium may not exist due to discontinuities in the reaction curves of the firms, which derive from the lack of quasi concavity of the profit functions (see Roberts and Sonnenschein (1977)) and that the (existence of) equilibrium depends upon the specific price normalization rule that is used (see Gabsewicz and Vial (1972) and Böhm (1994)).

A realistic price adjustment process for general equilibrium models would incorporate price-setting behaviour of economic agents and would fall within the literature discussed above. Notice that the equilibria in a general equilibrium model with price setting behaviour will in general not coincide with the Walrasian equilibria. As a first step in the direction of these more realistic price adjustment processes we will study some price adjustment processes in partial equilibrium models with monopolistic competition in Chapter 6.

2.3 The Model

In this section we introduce our model. In the first subsection we discuss the problem of finding an appropriate normalization rule and we suggest that for the multiplicative tâtonnement process normalizing prices on the simplex is the most natural normalization rule. In subsection 2.3.2 we specify our model and in subsection 2.3.3 we show some results obtained from numerical experiments.

2.3.1 Price normalization

The tâtonnement process (2.4) is a discrete n-dimensional dynamical system. From the homogeneity of the aggregate excess demand functions we have that if $\mathbf{p}^*$ is an equilibrium price vector then so is $\alpha\mathbf{p}^*$ for any $\alpha > 0$. Instead of an equilibrium price vector we have a *ray* of equilibrium price vectors. None of these equilibria can be locally stable since every neighbourhood in $\mathbb{R}^n_+$ of such an equilibrium point contains other equilibrium points (the tâtonnement process can however be system stable). The tâtonnement process then seems to be ill-defined.

The tâtonnement process can be restricted to a $n-1$ dimensional subspace of $\mathbb{R}^n$ by imposing an extra condition on the prices. Such a *price normalization rule* $g(\mathbf{p}) = c$, where c is a constant, defines a $n-1$ dimensional manifold in $\mathbb{R}^n_+$ and has to be chosen in such a way that the intersection of this manifold with any ray of equilibrium price vectors consists of exactly one point. This intersection point is the equilibrium price vector of the tâtonnement process restricted to this manifold. Some well-known price normalization rules are $g(\mathbf{p}) = p_k$ for some fixed k, $g(\mathbf{p}) = \sum_{j=1}^{n} p_j$ and $g(\mathbf{p}) = \sum_{j=1}^{n} p_j^2$, all with $c = 1$, respectively corresponding to one good being chosen as numeraire, prices on the unit simplex and prices on the unit sphere. There is no economic motivation to choose a particular price normalization rule. Prices are free and there is no particular reason why they should obey a certain condition. In special cases some normalization rule seems obvious (e.g. money as a numeraire in a monetary economy), but no normalization rule is appropriate *a priori*.

Having chosen a normalization rule it subsequently has to be implemented. Assume $g(\mathbf{p}) = c$ can be rewritten as $p_k = h_k(p_1, ..., p_{k-1}, p_{k+1}, ..., p_n)$ for some k. The tâtonnement process restricted to $g(\mathbf{p}) = c$ then becomes:

$$
\begin{aligned}
p_{j,t+1} &= p_{jt} + F_j\left(p_{jt}, z_j\left(p_{1t}, .., p_{k-1,t}, h_k\left(.\right), p_{k+1,t}, .., p_{nt}\right)\right), j \neq k, \\
p_{kt} &= h_k\left(p_{1t}, ..., p_{k-1,t}, p_{k+1,t}, ..., p_{nt}\right). \qquad (2.6)
\end{aligned}
$$

So the dimension of the dynamical system is reduced by one to $n-1$. In fact the k'th equation of motion has been deleted. From Walras' Law we have that if $n-1$ markets are in equilibrium the n'th market must also be in equilibrium, so an equilibrium price vector of (2.6) is also an equilibrium price vector of (2.4). The dynamics of the tâtonnement process however depend crucially on this reduction. Which equation is deleted obviously influences the dynamical system. The tâtonnement process with commodity 1 as numeraire may have different stability properties than the process with commodity n as numeraire. There is *"goods discrimination"* in the sense that the price of one, arbitrarily chosen, good is treated different than the prices of the other goods.

In some cases there is a nice solution to this last problem. If for some tâtonnement process $\mathbf{p}_{t+1} = f(\mathbf{p}_t)$ there exists a price normalization rule $g(.)$ such that $g(f_1(\mathbf{p}), f_2(\mathbf{p}), ..., f_n(\mathbf{p})) = g(p_1, p_2, ..., p_n)$ then the tâtonnement process f leaves the manifold defined by $g(.) = c$ *invariant*. In this case the n equations of motion of the tâtonnement process are not independent. In fact there is one redundant equation and any equation can be replaced by the normalization rule without affecting the dynamical system: the systems (2.6) and (2.4) are equivalent and no information is lost by imposing $g(.) = c$.

For the continuous time case such a normalization rule exists for the linear tâtonnement process $\frac{dp_i}{dt} = \lambda z_i(\mathbf{p}(t))$. We have:

$$\frac{d}{dt}\sum p_i^2(t) = 2\sum p_i(t)\frac{dp_i}{dt} = 2\lambda\sum p_i(t)\, z_i(\mathbf{p}(t)) = 0,$$

due to Walras' Law, so the linear tâtonnement process leaves the sphere invariant.

In the case of the discrete tâtonnement process a similar price normalization rule exists for the so-called multiplicative tâtonnement process

$$p_{j,t+1} = p_{jt}\left(1 + \lambda z_j(\mathbf{p}_t)\right) \qquad j = 1, ..., n, \qquad (2.7)$$

where λ is the speed of adjustment[4] and is assumed to be the same for all markets.[5] Summation of the prices gives (using Walras' Law):

$$\sum_{j=1}^{n} p_{j,t+1} = \sum_{j=1}^{n} p_{jt}\left(1 + \lambda z_j(\mathbf{p}_t)\right) = \sum_{j=1}^{n} p_{jt} + \lambda\sum_{j=1}^{n} p_{jt} z_j(\mathbf{p}_t) = \sum_{j=1}^{n} p_{jt}.$$

Hence the multiplicative tâtonnement process leaves the simplex invariant.

It can be argued that (2.7) is a reasonable way to model price adjustments. It states that the rate of change of the price of a good is proportional to its excess demand and it is homogeneous of degree one in prices. From now on we shall use (2.7) to study the dynamics of price adjustments.[6]

2.3.2 Model specifications

In order to analyze the global dynamics of our tâtonnement process we have to specify our model. We will study an economy with three commodities and three consumers that have CES (Constant Elasticity

[4] An alternative interpretation of λ is that it is a measure of the size of the economy.

[5] This is not a restrictive assumption since we can always normalize the unit of goods such that the speeds of adjustment are equal across markets.

[6] For the sake of simplicity we have not imposed a non-negativity constraint on the prices. If the speed of adjustment is not too large, say $\lambda < \overline{\lambda}$, then for initial price vectors chosen not too close to the border of the simplex the tâtonnement process stays on the simplex. If $\lambda > \overline{\lambda}$ then for almost all initial price vectors the price of some good becomes negative in finite time and the tâtonnement process breaks down. However, all kinds of interesting dynamics can be observed for $\lambda < \overline{\lambda}$.

of Substitution) utility functions, i.e. the utility function of consumer i is

$$U_i(x_{i1}, x_{i2}, x_{i3}) = \left(\sum_{j=1}^{3} \alpha_{ij} x_{ij}^{\rho}\right)^{\frac{1}{\rho}}, \qquad \rho < 0, \qquad i = 1,2,3. \tag{2.8}$$

The corresponding individual demand functions are

$$x_{ij}(p_1, p_2, p_3) = \frac{\left(\frac{\alpha_{ij}}{p_j}\right)^{\sigma}}{\sum_{k=1}^{3} p_k \left(\frac{\alpha_{ik}}{p_k}\right)^{\sigma}} \sum_{j=1}^{3} p_j w_{ij}, \qquad i,j = 1,2,3, \tag{2.9}$$

with $\sigma = \frac{1}{1-\rho}$ the (constant) elasticity of substitution. The individual demand functions are homogeneous of degree zero in prices $\mathbf{p}$ and in preference parameters α_{i1}, α_{i2} and α_{i3} and homogeneous of degree one in initial endowments $\mathbf{w}_i$. Under the assumption that $\alpha_{ii} > 0$ we can take, without loss of generality, $\alpha_{ii} = 1$.

We assume that the initial endowments are of the following form: $w_{ij} = 1$ if $i = j$ and $w_{ij} = 0$ if $i \neq j$. So each consumer supplies exactly one good and is the sole supplier of that good. This restriction is made for computational simplicity and does not influence the qualitative results. The multiplicative price adjustment process now becomes

$$\begin{aligned}
p_{1,t+1} &= p_{1t}\left(1 + \lambda\left[\sum_{i=1}^{3} \frac{p_{it}\left(\frac{\alpha_{i1}}{p_{1t}}\right)^{\sigma}}{\sum_{j=1}^{3} p_{jt}\left(\frac{\alpha_{ij}}{p_{jt}}\right)^{\sigma}} - 1\right]\right), \\
p_{2,t+1} &= p_{2t}\left(1 + \lambda\left[\sum_{i=1}^{3} \frac{p_{it}\left(\frac{\alpha_{i2}}{p_{2t}}\right)^{\sigma}}{\sum_{j=1}^{3} p_{jt}\left(\frac{\alpha_{ij}}{p_{jt}}\right)^{\sigma}} - 1\right]\right), \\
p_{3,t+1} &= p_{3t}\left(1 + \lambda\left[\sum_{i=1}^{3} \frac{p_{it}\left(\frac{\alpha_{i3}}{p_{3t}}\right)^{\sigma}}{\sum_{j=1}^{3} p_{jt}\left(\frac{\alpha_{ij}}{p_{jt}}\right)^{\sigma}} - 1\right]\right).
\end{aligned} \tag{2.10}$$

This is the dynamical system under study. The parameters of the model are α_{ij}, σ and λ. We will study the behaviour of the dynamical system

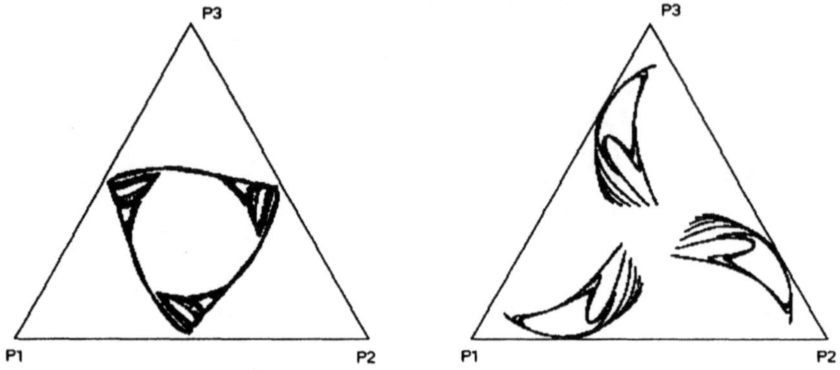

Figure 2.1: Attractors of the tâtonnement process with rotational symmetry. a) $\alpha_{12} = \alpha_{23} = \alpha_{31} = 1$, $\alpha_{13} = \alpha_{21} = \alpha_{32} = \frac{1}{2}$, $\sigma = \frac{1}{2}$ and $\lambda = 4.95$. b) $\alpha_{12} = \alpha_{23} = \alpha_{31} = 1$, $\alpha_{13} = \alpha_{21} = \alpha_{32} = \frac{1}{4}$, $\sigma = \frac{1}{4}$ and $\lambda = 8.45$.

for given preference parameters α_{ij} and given elasticity of substitution σ, as the speed of adjustment λ increases. We focus on $\alpha_{ij} \in (0, 1]$ and $\sigma \in (0, 1]$. For most of the specifications of our model we will study, the equilibrium price vector is unique. Section 2.8 briefly deals with the case of multiple equilibria.

2.3.3 Some typical numerical simulations

Before analyzing our tâtonnement process in more detail we present some numerical results. We are interested in the long run behaviour of our dynamical system: do prices settle down to an equilibrium price vector or is the discrete tâtonnement process unstable and can prices fluctuate forever? To describe the long run behaviour of the dynamical system we use the following definition (Guckenheimer and Holmes (1983), p.36).

Definition 2.1 *A compact set* $\mathbf{A}$ *is called an* ***attractor*** *of a dynamical system* f *if:*

1) $\mathbf{A}$ *is invariant with respect to* $f : f(\mathbf{A}) \subset \mathbf{A}$.

2) $\mathbf{A}$ *is attracting, that is there is an open neighbourhood* $\mathbf{U}$ *of* $\mathbf{A}$ *such that the trajectory of every initial state* $\mathbf{p} \in \mathbf{U}$ *converges to* $\mathbf{A}$.

3) $\mathbf{A}$ *has a dense orbit.*

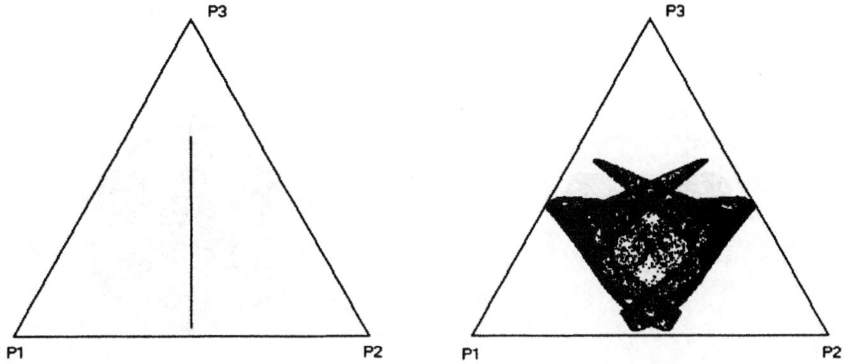

Figure 2.2: Attractors of the tâtonnement process with reflectional symmetry. a) $\alpha_{12} = \alpha_{21} = \frac{1}{4}$, $\alpha_{13} = \alpha_{23} = \alpha_{31} = \alpha_{32} = \frac{1}{2}$, $\sigma = \frac{1}{10}$ and $\lambda = 32$. b) $\alpha_{12} = \alpha_{21} = \frac{3}{4}$, $\alpha_{13} = \alpha_{23} = \alpha_{31} = \alpha_{32} = \frac{1}{2}$, $\sigma = \frac{1}{10}$ and $\lambda = 31$.

The simplest example of an attractor is a locally stable fixed point. Another simple example is an attracting periodic orbit. In that case we have $f^n(\mathbf{p}) = \mathbf{p}$ and $f^k(\mathbf{p}) \neq \mathbf{p},\ k < n$. However, more complicated *quasi-periodic* attractors or attractors with a fractal structure can arise. This last type of attractors is sometimes called *strange* or *chaotic* since the dynamics on these attractors may seem unpredictable and shows *sensitive dependence on initial conditions.*

Some typical examples of attractors are shown in Figures 2.1-2.3.[7] All these attractors were obtained by iterating (2.10) for different values of the preference parameters α_{ij}, the elasticity of substitution σ and the speed of adjustment λ.[8]

Notice that the attractors have a certain *symmetry.* The pictures in Figure 2.1 have a *rotation symmetry*: rotating the pictures over an angle of $\frac{2}{3}\pi$ results in exactly the same pictures. The pictures in Figure 2.2 have a *reflection symmetry*: reflection of the pictures in Figure 2.2 in the line with $p_1 = p_2$ results in the same pictures. The attractors in Figure 2.3 have a rotation as well as a reflection symmetry: the pictures are

[7] To plot the pictures in two dimensions we have transformed the simplex in $\mathbb{R}^3_+$ into a triangle in $\mathbb{R}^2$. This transformation is also used in the proof of proposition 2.2.

[8] All figures of attractors in this paper (with the exception of periodic orbits) were obtained by iterating the dynamical system 6.000 times and plotting the last 5.000 points.

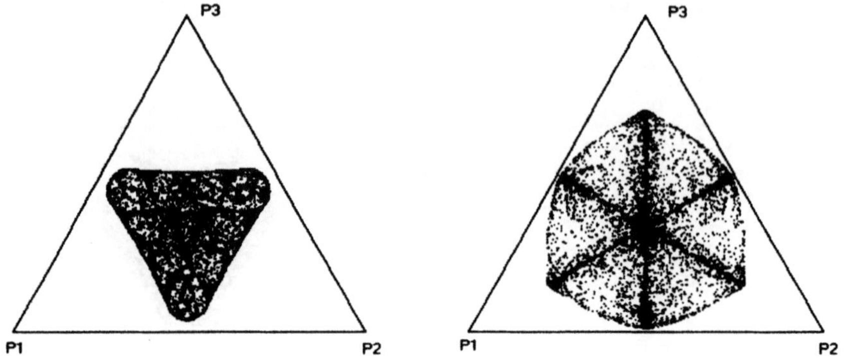

Figure 2.3: Strange attractors of the tâtonnement process with cyclical and reflectional symmetry. a) $\alpha_{ij} = 1$, all i, j, $\sigma = \frac{1}{10}$ and $\lambda = 26$. b) $\alpha_{12} = \alpha_{13} = \alpha_{21} = \alpha_{23} = \alpha_{31} = \alpha_{32} = \frac{1}{10}$, $\sigma = \frac{1}{10}$ and $\lambda = 95$.

invariant both to a rotation over $\frac{2}{3}\pi$ and to a reflection in the line with $p_1 = p_2$.

The prices move erratically across these attractors. The attractors show the long run behaviour of the prices of the tâtonnement process. From these simulations it is clear that for many parameter settings the tâtonnement process will not settle down to an equilibrium price vector but will keep on fluctuating forever. In the subsequent sections we will investigate bifurcation routes leading to these strange attractors. Symmetry will play an important role in characterizing these bifurcation routes.

2.4 Symmetry

A priori the number of independent parameters of the tâtonnement process with preferences given by (2.8) is 17 (α_{ij},w_{ij},λ and σ). To study this model in a transparent manner we have to reduce this number. A first reduction was achieved by taking endowments $w_{ii} = 1$ and $w_{ij} = 0$ for $i \neq j$. This leaves us with 8 parameters. The simplex as such has both a rotation and a reflection symmetry. Therefore it seems to be natural to start a systematic investigation of the global dynamics by firstly focussing on symmetric endowments and preferences, to be defined below. In Section 2.7 we discuss properties of close-by asymmetric cases. There is a large literature on symmetry in dynamical systems (see for example Golubitsky, Stewart and Schaeffer (1988)). We will apply concepts and results from this literature to our multiplicative tâtonnement process.

Consider the following dynamical system on the two-dimensional simplex

$$\mathbf{p}_{t+1} = f(\mathbf{p}_t), \qquad \mathbf{p}_t \in \mathbf{S}^2 = \left\{ \mathbf{p} \in I\!R^3_+ \mid \sum_{j=1}^{3} p_j = 3 \right\}. \tag{2.11}$$

A linear transformation $M : I\!R^3 \to I\!R^3$ is called a symmetry of f if M and f commute, that is, if

$$M \circ f = f \circ M. \tag{2.12}$$

Let M and N be symmetries of f and let $\mathbf{p}^*$ be an equilibrium price vector of f. Then we have the following properties

- $M \circ N$ is a symmetry of f, $M \circ N \circ f = M \circ f \circ N = f \circ M \circ N$,
- M and N are symmetries of all iterates of f, $M \circ f^2 = f \circ M \circ f = f^2 \circ M$, etc.,
- $M\mathbf{p}^*$ is an equilibrium price vector of f, $f(M\mathbf{p}^*) = Mf(\mathbf{p}^*) = M\mathbf{p}^*$.

The simplex is invariant with respect to any permutation of prices, so the group of symmetries that is interesting to study for the tâtonnement process on the simplex is the group of permutation matrices $\mathbf{D}_3 = \{M_{123}, M_{132}, M_{321}, M_{213}, M_{312}, M_{231} \mid M_{ijk} \in I\!R^{3\times 3}_+\}$, with $M_{ijk} \begin{pmatrix} p_1 \\ p_2 \\ p_3 \end{pmatrix} = \begin{pmatrix} p_i \\ p_j \\ p_k \end{pmatrix}$. Of course $M_{123} = I$ is a trivial symmetry of all dynamical systems. Apart from I and $\mathbf{D}_3$ there are four different subgroups of $\mathbf{D}_3$, $\{I, M_{132}\}, \{I, M_{321}\}$, $\{I, M_{213}\}$ and $\{I, M_{312}, M_{231}\}$. The first three of these are so-called $\mathbf{Z}_2$ symmetries: they correspond to a permutation of two of the three prices or equivalently a reflection in one of the symmetry axes of the simplex. These symmetry axes are: $l_1 = \{\mathbf{p} \in \mathbf{S}^2 \mid p_2 = p_3\}, l_2 = \{\mathbf{p} \in \mathbf{S}^2 \mid p_1 = p_3\}$ and $l_3 = \{\mathbf{p} \in \mathbf{S}^2 \mid p_1 = p_2\}$. The last subgroup is the $\mathbf{R}_3$ symmetry and corresponds to a cyclical permutation of the three prices, or equivalently a rotation over $\frac{2}{3}\pi$.

We can now define the symmetry group of a price adjustment process f on the simplex as follows: $\Gamma_f = \{M \in \mathbf{D}_3 | M \circ f = f \circ M\}$. By choosing suitable preferences and endowments we will study a price adjustment process with a $\mathbf{Z}_2$ symmetry, one with a $\mathbf{R}_3$ symmetry and a price adjustment process with both symmetries, which is equivalent to symmetry group $\mathbf{D}_3$. So we study tâtonnement processes with three different symmetry groups: $\Gamma_f = \{I, M_{312}, M_{231}\}$, $\Gamma_f = \{I, M_{213}\}$ and $\Gamma_f = \mathbf{D}_3$. These symmetry groups correspond respectively to tâtonnement processes that are invariant with respect to a cyclic, a reflective or all possible permutations of prices.

Definition 2.2 *The* ***fixed point subspace*** $Fix\,(M)$ *of a symmetry* M *is the set*

$$Fix\,(M) = \{\mathbf{p} \in \mathbf{S}^2 \mid M\mathbf{p} = \mathbf{p}\}.$$

This fixed point subspace plays an important role in the bifurcations of the equilibrium price vector in a symmetric tâtonnement process. $Fix\,(M)$ is a linear subspace that consists of all points of the simplex that are invariant under symmetry M. $Fix\,(M)$ is a line if M is a reflectional symmetry and $Fix\,(M)$ is a point if M is a rotational symmetry, for example $Fix\,(M_{213}) = l_3$ and $Fix\,(M_{312}) = (1,1,1)$. It can be easily seen that if $M \in \Gamma_f$ then f leaves $Fix\,(M)$ invariant, since for $\mathbf{p} \in Fix\,(M)$ we have

$$f(\mathbf{p}) = f(M\mathbf{p}) = Mf(\mathbf{p}),$$

so $f(\mathbf{p}) \in Fix\,(M)$. This means that a *nonlinear* dynamical system with a symmetry has a *linear* invariant space. This property will prove to be very useful when we study the bifurcations of tâtonnement processes with a reflectional symmetry. This property also implies that $(1,1,1)$ is an equilibrium price vector of all tâtonnement processes that have a rotational symmetry, since $Fix\,(M_{312}) = (1,1,1)$.

Now we want to impose symmetry on our economy with CES-utility functions. The following proposition establishes how we can construct a tâtonnement process with a rotational and a reflectional symmetry, respectively, if there are three agents and three commodities.

Proposition 2.1 *Let $U, V: \mathbb{R}^3_+ \to \mathbb{R}$ be utility functions, $\mathbf{p} = (p_1, p_2, p_3)'$ a vector of prices and $\mathbf{w} = (w_1, w_2, w_3)'$, $\mathbf{v} = (v_1, v_1, v_3)'$ vectors of initial endowments. Consider the two permutation matrices M_{231} and M_{213}. Take $V(.)$ such that $V(M_{213}\mathbf{x}) = V(\mathbf{x})$.*

- *Consider an economy with three consumers with utility functions $U_1(\mathbf{x}) = U(\mathbf{x})$, $U_2(\mathbf{x}) = U(M_{231}\mathbf{x})$ and $U_3(\mathbf{x}) = U(M_{231}^2\mathbf{x})$ and vectors of initial endowments $\mathbf{w}_1 = \mathbf{w}$, $\mathbf{w}_2 = M_{231}'\mathbf{w}$ and $\mathbf{w}_3 = (M_{231}')^2\mathbf{w}$, respectively. Then the corresponding multiplicative tâtonnement process has rotation symmetry M_{231}.*

- *Consider an economy with three consumers with utility functions $U_1(\mathbf{x}) = U(\mathbf{x})$, $U_2(\mathbf{x}) = U(M_{213}\mathbf{x})$ and $U_3(\mathbf{x}) = V(\mathbf{x})$ and vectors of initial endowments $\mathbf{w}_1 = \mathbf{w}$, $\mathbf{w}_2 = M_{213}\mathbf{w}$ and $\mathbf{w}_3 = \mathbf{v}$, respectively. Then the corresponding multiplicative tâtonnement process has reflectional symmetry M_{213}.*

Proof. Let the utility maximizing consumption bundle be given by

$$\mathbf{x} = \mathbf{d}^U(\mathbf{p}, \mathbf{w}) = \arg\max_{\mathbf{x} \in \mathbb{R}^m_+} \{U(\mathbf{x}) \mid \mathbf{p}'\mathbf{x} \leq \mathbf{p}'\mathbf{w}\}.$$

Consider the problem of maximizing $U(M\mathbf{x})$ subject to the budget constraint $\mathbf{p}'\mathbf{x} \leq \mathbf{p}'\left(M'\mathbf{w}\right)$, where $M \in \{M_{231}, M_{213}\}$. Since $M^{-1} = M'$, this problem is equivalent to maximizing $U(M\mathbf{x})$ subject to $(M\mathbf{p})' M\mathbf{x} \leq (M\mathbf{p})'\mathbf{w}$. The solution to this last problem is easily seen to be

$$\mathbf{x} = M'\mathbf{d}(M\mathbf{p}, \mathbf{w}).$$

- First consider $M = M_{231}$. Notice that $M_{231}^2 = M_{231}'$ and $M_{231}^3 = I$. The individual demand functions now become $\mathbf{x}_{k+1} = (M_{231}')^k \mathbf{d}^U(M_{231}^k\mathbf{p}, \mathbf{w})$, with $k = 0, 1, 2$. The aggregate excess demand functions $\mathbf{z}(\mathbf{p}) = \sum_{i=1}^3 \mathbf{x}_i - \sum_{i=1}^3 \mathbf{w}_i$ can be written as $\mathbf{z}(\mathbf{p}, \mathbf{w}) =$

$$\begin{pmatrix} d_1^U(\mathbf{p}, \mathbf{w}) + d_2^U(M_{231}^2\mathbf{p}, \mathbf{w}) + d_3^U(M_{231}\mathbf{p}, \mathbf{w}) - w_1 - w_2 - w_3 \\ d_1^U(M_{231}\mathbf{p}, \mathbf{w}) + d_2^U(\mathbf{p}, \mathbf{w}) + d_3^U(M_{231}^2\mathbf{p}, \mathbf{w}) - w_1 - w_2 - w_3 \\ d_1^U(M_{231}^2\mathbf{p}, \mathbf{w}) + d_2^U(M_{231}\mathbf{p}, \mathbf{w}) + d_3^U(\mathbf{p}, \mathbf{w}) - w_1 - w_2 - w_3 \end{pmatrix}.$$

It can be easily checked that we have $\mathbf{z}(M_{231}\mathbf{p}, \mathbf{w}) = M_{231}\mathbf{z}(\mathbf{p}, \mathbf{w})$. This means that M_{231} is a symmetry of the aggregate excess demand functions.

- Now consider $M = M_{213}$. The demand functions are respectively $\mathbf{x}_1 = \mathbf{d}^U(\mathbf{p},\mathbf{w})$, $\mathbf{x}_2 = M_{213}\mathbf{d}^U(M_{213}\mathbf{p},\mathbf{w})$ and $\mathbf{x}_3 = \mathbf{d}^V(\mathbf{p},\mathbf{v})$ with $d_2^V(\mathbf{p},\mathbf{v}) = d_1^V(M_{213}\mathbf{p},\mathbf{v})$ and $d_3^V(M_{213}\mathbf{p},\mathbf{v}) = d_3^V(\mathbf{p},\mathbf{v})$. The aggregate excess demand functions become $\mathbf{z}(\mathbf{p},\mathbf{w}) =$

$$\begin{pmatrix} d_1^U(\mathbf{p},\mathbf{w}) + d_2^U(M_{213}\mathbf{p},\mathbf{w}) + d_1^V(\mathbf{p},\mathbf{v}) - w_1 - w_2 - v_1 \\ d_1^U(M_{213}\mathbf{p},\mathbf{w}) + d_2^U(\mathbf{p},\mathbf{w}) + d_1^V(M_{213}\mathbf{p},\mathbf{v}) - w_1 - w_2 - v_1 \\ d_3^U(\mathbf{p},\mathbf{w}) + d_3^U(M_{213}\mathbf{p},\mathbf{w}) + d_3^V(\mathbf{p},\mathbf{v}) - 2w_3 - v_3 \end{pmatrix}.$$

We have $M_{213}\mathbf{z}(\mathbf{p},\mathbf{w}) = \mathbf{z}(M_{213}\mathbf{p},\mathbf{w})$.

Now if the aggregate demand functions have symmetry M_{231} then for the multiplicative price adjustment process we have

$$\begin{pmatrix} f_1(M_{231}\mathbf{p}) \\ f_2(M_{231}\mathbf{p}) \\ f_3(M_{231}\mathbf{p}) \end{pmatrix} = \begin{pmatrix} p_{2t}(1+\lambda z_1(M_{231}\mathbf{p})) \\ p_{3t}(1+\lambda z_2(M_{231}\mathbf{p})) \\ p_{1t}(1+\lambda z_3(M_{231}\mathbf{p})) \end{pmatrix}$$
$$= \begin{pmatrix} p_{2t}(1+\lambda z_2(\mathbf{p})) \\ p_{3t}(1+\lambda z_3(\mathbf{p})) \\ p_{1t}(1+\lambda z_1(\mathbf{p})) \end{pmatrix} = M_{231}f(\mathbf{p})$$

and if the aggregate demand functions have symmetry M_{213} then for the multiplicative price adjustment process we have

$$\begin{pmatrix} f_1(M_{213}\mathbf{p}) \\ f_2(M_{213}\mathbf{p}) \\ f_3(M_{213}\mathbf{p}) \end{pmatrix} = \begin{pmatrix} p_{2t}(1+\lambda z_1(M_{213}\mathbf{p})) \\ p_{1t}(1+\lambda z_2(M_{213}\mathbf{p})) \\ p_{3t}(1+\lambda z_3(M_{213}\mathbf{p})) \end{pmatrix}$$
$$= \begin{pmatrix} p_{2t}(1+\lambda z_2(\mathbf{p})) \\ p_{1t}(1+\lambda z_1(\mathbf{p})) \\ p_{3t}(1+\lambda z_3(\mathbf{p})) \end{pmatrix} = M_{213}f(\mathbf{p}). \blacksquare$$

This result gives us a way to impose symmetry on our price adjustment process with CES utility functions. Let $A = (\alpha_{ij})$ be the matrix of preferences. We then have

Corollary 2.1 *For an economy with three commodities and three agents with CES utility functions and initial endowments $w_{ii} = 1$ and $w_{ij} = 0$ for $i \neq j$ we have that the following types of preference matrices*

$$A^{\mathbf{R}_3} = \begin{pmatrix} 1 & \alpha & \beta \\ \beta & 1 & \alpha \\ \alpha & \beta & 1 \end{pmatrix}, A^{\mathbf{Z}_2} = \begin{pmatrix} 1 & \alpha & \beta \\ \alpha & 1 & \beta \\ \gamma & \gamma & 1 \end{pmatrix}, A^{\mathbf{D}_3} = \begin{pmatrix} 1 & \alpha & \alpha \\ \alpha & 1 & \alpha \\ \alpha & \alpha & 1 \end{pmatrix}, \tag{2.13}$$

induce a rotational, a reflectional and both kinds of symmetries of the multiplicative price adjustment process, respectively.

The attractors shown in Section 2.3.3 were obtained by using the matrices (2.13). The attractors in Figures 2.1-2.3 obviously exhibit the same symmetry as the corresponding tâtonnement processes. Rotating the pictures in Figure 2.1 over $\frac{2}{3}\pi$ results in exactly the same pictures and reflection of the attractors in Figure 2.2 in the line with $p_1 = p_2$ leaves the attractors invariant. For the pictures in Figure 2.3 any permutation of the prices can be executed without changing the picture. The symmetry of the tâtonnement process seems to determine the symmetry of the resulting attractor.

If f has symmetry M and $\mathbf{p}_0$ converges to an attractor $\mathbf{A}$ then obviously $M\mathbf{p}_0$ converges to an attractor $M\mathbf{A}$. If they converge to the same attractor, that is if $M\mathbf{A} = \mathbf{A}$, then this attractor can be said to have symmetry M. So we can define the symmetry group of an attractor of a dynamical system with symmetry group Γ_f as $\Sigma_{\mathbf{A}} = \{M \in \Gamma_f \mid M\mathbf{A} = \mathbf{A}\}$. For the attractors of Section 2.3.3 we have $\Sigma_{\mathbf{A}} = \Gamma_f$. This might suggest that the attractor always has the same symmetry group as the dynamical system, but in general that need not be the case. We can have $M\mathbf{A} \neq \mathbf{A}$ and then we must have coexistence of attractors. These coexisting attractors are related by symmetry and are called *conjugate* attractors. In the sequel we shall encounter symmetry breaking bifurcations, where a single attractor breaks up into multiple coexisting attractors with a smaller symmetry group as a parameter changes, and symmetry increasing bifurcations where coexisting conjugate attractors merge into one single (symmetric) attractor.

2.5 Local bifurcation analysis

In this section we study which local bifurcations can occur in our tâtonnement process. We thereby focus on the symmetric equilibrium price vector. In subsection 2.5.1 we show that the symmetry of the tâtonnement process determines the form of the Jacobian matrix. In subsection 2.5.2-2.5.4 we study the implications of this result for the generic[9] bifurcations that occur for the tâtonnement processes with different symmetries.

[9]Here generic means with respect to the set of dynamical systems with a given symmetry group.

2.5.1 Symmetry and the Jacobian matrix

To study the local bifurcations of the equilibrium price vector that can occur in our tâtonnement process we have to look at the eigenvalues μ_i of the Jacobian matrix $\mathbf{J} = \left(\frac{\partial f_i}{\partial p_j}(\mathbf{p})\right)$ evaluated at the equilibrium price vector $\mathbf{p}^*$. These eigenvalues depend on the parameters of the tâtonnement process, so we have $\mu_i = \mu_i(\lambda, \sigma, A)$ where A is the matrix with preference parameters. The equilibrium price vector $\mathbf{p}^*$ is locally stable if all eigenvalues lie inside the unit circle and $\mathbf{p}^*$ is unstable when at least one eigenvalue lies outside the unit circle. If an eigenvalue crosses the unit circle, as a parameter varies, a bifurcation occurs. At such a bifurcation the behaviour of the tâtonnement process changes. There are three different ways in which an eigenvalue can cross the unit circle. At the bifurcation value an eigenvalue can be real and equal to $+1$, real and equal to -1 or two eigenvalues can be complex conjugates with absolute value 1. These three different cases correspond to different kinds of bifurcations.

First we show that symmetry induces a special form of the Jacobian matrix evaluated in the symmetric equilibrium price vector.

Proposition 2.2 *The Jacobian of the two-dimensional tâtonnement process evaluated at a symmetric equilibrium has eigenvalues that are*

- *real if the tâtonnement process has a* $\mathbf{Z}_2$ *symmetry. In this case one eigenvector lies in the fixed point subspace of the* $\mathbf{Z}_2$ *symmetry and the other eigenvector lies perpendicular to this fixed point subspace,*
- *complex conjugates if the tâtonnement process has a* $\mathbf{R}_3$ *symmetry,*
- *real and equal to each other if the tâtonnement process has both symmetries.*

Proof. Without loss of generality we can apply a linear transformation to the prices to transform the simplex in $I\!R^3_+$ to a triangle in the plane. Let $\mathbf{q} = \mathcal{T}\mathbf{p}$ with

$$\mathcal{T} = \begin{pmatrix} \frac{1}{2}\sqrt{3} & -\frac{1}{2}\sqrt{3} & 0 \\ -\frac{1}{2} & -\frac{1}{2} & 1 \\ 1 & 1 & 1 \end{pmatrix} \tag{2.14}$$

The price adjustment process in terms of these new variables becomes $\mathbf{q}_{t+1} = \mathcal{T} \circ f \circ \mathcal{T}^{-1}(\mathbf{q}_t) = h(\mathbf{q}_t)$. Since we have $q_{3t} = \sum_{i=1}^{3} p_{it} = 3$, for all t, this is equivalent to

$$\begin{pmatrix} q_{1,t+1} \\ q_{2,t+1} \end{pmatrix} = \begin{pmatrix} h_1(q_{1t}, q_{2t}, 3) \\ h_2(q_{1t}, q_{2t}, 3) \end{pmatrix}. \tag{2.15}$$

Notice that $\mathcal{T}$ transforms the centre of the simplex to the origin and the symmetry axis l_3 to the q_2 axis. We have $\mathbf{J}_h = \mathcal{T}\mathbf{J}_f\mathcal{T}^{-1}$ and the eigenvalues of $\mathbf{J}_h$ are equal to the eigenvalues of $\mathbf{J}_f$. The $\mathbf{Z}_2$ and $\mathbf{R}_3$ symmetries can be represented respectively by

$$M_{\mathbf{z}} = \begin{pmatrix} -1 & 0 \\ 0 & 1 \end{pmatrix}, \qquad M_{\mathbf{r}} = \begin{pmatrix} -\frac{1}{2} & \frac{1}{2}\sqrt{3} \\ -\frac{1}{2}\sqrt{3} & -\frac{1}{2} \end{pmatrix}. \tag{2.16}$$

$M_{\mathbf{z}}$ corresponds to a reflection in the q_2 axis and $M_{\mathbf{r}}$ corresponds to a rotation over $\frac{2}{3}\pi$. With this transformation of variables it is easy to establish our result. Let M be a symmetry of h. Then by definition we have $Mh(\mathbf{q}) = h(M\mathbf{q})$. Differentiating both sides with respect to $\mathbf{q}$ gives

$$M\frac{\partial h}{\partial \mathbf{q}'}(\mathbf{q}) = \frac{\partial h}{\partial \mathbf{q}'}(M\mathbf{q})\, M.$$

Now if $\mathbf{q}^*$ is symmetric (that is $M\mathbf{q}^* = \mathbf{q}^*$) we have

$$M\mathbf{J}_h^* = \mathbf{J}_h^* M.$$

So the Jacobian evaluated at the symmetric equilibrium commutes with the symmetries of the dynamical system. An easy calculation shows that the only 2×2 matrices commuting with $M_{\mathbf{z}}$ and with $M_{\mathbf{r}}$ are $\mathbf{J}_{\mathbf{z}}$ and $\mathbf{J}_{\mathbf{r}}$, respectively with

$$\mathbf{J}_{\mathbf{z}} = \begin{pmatrix} a & 0 \\ 0 & b \end{pmatrix}, \qquad \mathbf{J}_{\mathbf{r}} = \begin{pmatrix} a & b \\ -b & a \end{pmatrix}.$$

The eigenvalues of $\mathbf{J}_{\mathbf{z}}$ are a and b with eigenvectors $\nu_a = (1,0)'$ and $\nu_b = (0,1)'$ respectively, the eigenvalues of $\mathbf{J}_{\mathbf{r}}$ are $a \pm bi$ and therefore complex as long as $b \neq 0$. Note that if the dynamical system has both symmetries we have

$$\mathbf{J}_{\mathbf{D}_3} = \begin{pmatrix} a & 0 \\ 0 & a \end{pmatrix} = aI.$$

The statements of the proposition follow. ■

From this proposition it follows that the symmetry of the tâtonnement process determines which primary bifurcation occurs as the symmetric equilibrium price vector loses stability.

We now derive the eigenvalues of the Jacobian evaluated at the symmetric equilibrium for each symmetry. The Jacobian of our tâtonnement process (2.7) evaluated in the equilibrium price vector $\mathbf{p}^*$ is

$$\mathbf{J}^* = I + \lambda \begin{pmatrix} p_1^* z_{11}(\mathbf{p}^*) & p_1^* z_{12}(\mathbf{p}^*) & p_1^* z_{13}(\mathbf{p}^*) \\ p_2^* z_{21}(\mathbf{p}^*) & p_2^* z_{22}(\mathbf{p}^*) & p_2^* z_{23}(\mathbf{p}^*) \\ p_3^* z_{31}(\mathbf{p}^*) & p_3^* z_{32}(\mathbf{p}^*) & p_3^* z_{33}(\mathbf{p}^*) \end{pmatrix}, \tag{2.17}$$

where $z_{ij}(\mathbf{p}) \equiv \frac{\partial z_i(\mathbf{p})}{\partial p_j}$. Notice that the matrix $(p_i^* z_{ij}(\mathbf{p}^*))_{i,j=1,2,3}$ has linear dependent rows ($\sum_{i=1}^3 p_i^* z_{ij}(\mathbf{p}^*) = -z_i(\mathbf{p}^*) = 0$ from Walras' Law) which implies that one eigenvalue of $\mathbf{J}^*$ is equal to 1, for all possible parameter values. This unit eigenvalue corresponds to the fact that every simplex is invariant under the multiplicative tâtonnement process. Since we are only interested in the motion of the dynamical system on the simplex we can safely ignore this eigenvalue. The other two eigenvalues are

$$\begin{aligned} \mu_{1,2} &= 1 + \frac{1}{2}\lambda\left\{p_1^*(z_{11} - z_{13}) + p_2^*(z_{22} - z_{23}) \pm \sqrt{D}\right\} \\ D &= \left[p_1^*(z_{11} - z_{13}) - p_2^*(z_{22} - z_{23})\right]^2 + 4p_1^*p_2^*(z_{12} - z_{13})(z_{21} - z_{23}) \end{aligned} \tag{2.18}$$

Under symmetry the expressions for these eigenvalues simplify considerably. These expressions are summarized in the following table

Γ_f	$\mu_{1,2}$
$\mathbf{R}_3$	$1 + \frac{1}{2}\lambda\left\{3z_{11}(\mathbf{p}^*) \pm (z_{12}(\mathbf{p}^*) - z_{13}(\mathbf{p}^*))\sqrt{3}i\right\}$
$\mathbf{Z}_2$	$1 + \lambda p_1^*\left\{z_{11}(\mathbf{p}^*) - z_{13}(\mathbf{p}^*) \pm (z_{12}(\mathbf{p}^*) - z_{13}(\mathbf{p}^*))\right\}$
$\mathbf{D}_3$	$1 + \frac{3}{2}\lambda z_{11}(\mathbf{p}^*)$

(2.19)

Notice that if $\Gamma_f = \mathbf{R}_3$ or $\Gamma_f = \mathbf{D}_3$ and $z_{11}(\mathbf{p}^*) > 0$ both eigenvalues lay outside the unit circle for all $\lambda > 0$. Hence, in that case the equilibrium price vector is unstable for all positive values of λ.

In the following subsections we analyze which bifurcations occur for each of the symmetry groups, as the speed of adjustment λ increases.

2.5.2 Rotational symmetry and the Hopf bifurcation

In the previous subsection we found that the eigenvalues of the Jacobian of the tâtonnement process with rotational symmetry (and no other symmetry) evaluated at the symmetric equilibrium price vector $\mathbf{p}^* = (1,1,1)$ are complex conjugates. The equilibrium price vector loses stability if these eigenvalues cross the unit circle. Let λ^{bif} denote the value of λ such that $\left|\mu_1\left(\lambda^{bif}\right)\right| = \left|\mu_2\left(\lambda^{bif}\right)\right| = 1$. We then have the following.

Proposition 2.3 *Assume that the tâtonnement process with symmetry group $\Gamma_f = \{I, M_{312}\}$ has a stable symmetric equilibrium price vector for $\lambda > 0$ small enough and that $z_{12}(\mathbf{p}^*) \neq z_{13}(\mathbf{p}^*)$. Furthermore assume that $\left(\mu_i\left(\lambda^{bif}\right)\right)^r \neq 1$ for $r = 1,2,3,4$. Then a Hopf bifurcation occurs at*

$$\lambda^{bif} = -\frac{4z_{11}(\mathbf{p}^*)}{3\left[z_{11}(\mathbf{p}^*)\right]^2 + \left[z_{12}(\mathbf{p}^*) - z_{13}(\mathbf{p}^*)\right]^2}$$

For λ close to λ^{bif} there exists an invariant closed curve.

Numerical observations suggest the following: for $\lambda < \lambda^{bif}$ the equilibrium price vector $\mathbf{p}^* = (1,1,1)$ is locally stable. For $\lambda > \lambda^{bif}$ the equilibrium price vector is unstable and there is an attracting invariant closed curve. This closed curve has the same symmetry group as the tâtonnement process. Figure 2.4.a) shows an example of such an attracting invariant closed curve obtained after a Hopf bifurcation.

2.5.3 Reflection symmetry and the period-doubling bifurcation

Consider the tâtonnement process with symmetry group $\Gamma_f = \{I, M_{213}\}$. The fixed point subspace of this $\mathbf{Z}_2$ symmetry is the symmetry axis l_3. From proposition 2.2 we found that the eigenvalues of the Jacobian of this dynamical system evaluated in a point in the fixed point subspace are real and have eigenvectors that lie perpendicular to the fixed point subspace and in the fixed point subspace, respectively.

The equilibrium price vector can only lose stability when the largest (in absolute value) eigenvalue goes through ± 1. If an eigenvalue goes through $+1$ a *saddle-node*, *pitchfork* or *transcritical bifurcation* occurs (see e.g. Guckenheimer and Holmes, (1983)). All these bifurcations in some way deal with the multiplicity of equilibria. Since the equilibrium price vectors of the tâtonnement process correspond to the zeros of the system of aggregate excess demand functions, the location and multiplicity of these equilibrium price vectors is independent of the speed of adjustment λ. Therefore we know that these bifurcations cannot occur in our model as λ increases. These bifurcations might occur, however, when another parameter is varied (see Section 2.8). When one of the eigenvalues goes through -1 a *flip- or period doubling bifurcation* occurs: at this bifurcation a period two orbit emerges. If this period doubling bifurcation occurs at $\lambda = \lambda^{bif}$ then we have the following scenario. For $\lambda < \lambda^{bif}$ the equilibrium price vector is stable and for $\lambda > \lambda^{bif}$ the equilibrium price vector is unstable. For λ close to λ^{bif} a period two orbit exists. Numerical observations suggest that this period two orbit exists for $\lambda > \lambda^{bif}$ and that it is stable.

Proposition 2.4 *Assume that the equilibrium price vector of the tâtonnement process with symmetry group* $\Gamma_f = \{I, M_{213}\}$ *is stable for* $\lambda > 0$ *small enough and that* $z_{12}(\mathbf{p}^*) \neq z_{13}(\mathbf{p}^*)$. *Then we have*

- *the equilibrium loses stability through a period doubling bifurcation at*

$$\lambda^{bif} = \frac{2}{p_1^*} \min \left\{ \frac{-1}{z_{11}(\mathbf{p}^*) - z_{12}(\mathbf{p}^*)}, \frac{-1}{z_{11}(\mathbf{p}^*) + z_{12}(\mathbf{p}^*) - 2z_{13}(\mathbf{p}^*)} \right\} \tag{2.20}$$

- *the resulting period two orbit has the same symmetry group as the tâtonnement process. The period two orbit lies in* l_3 *if* $z_{12}(\mathbf{p}^*) < z_{13}(\mathbf{p}^*)$ *and it lies off but symmetric with respect to* l_3 *when* $z_{12}(\mathbf{p}^*) > z_{13}(\mathbf{p}^*)$.[10]

Proof. The eigenvalues are

$$\begin{aligned}
\mu_1 &= 1 + \lambda p_1^* \left(z_{11}(\mathbf{p}^*) - z_{12}(\mathbf{p}^*)\right), \\
\mu_2 &= 1 + \lambda p_1^* \left(z_{11}(\mathbf{p}^*) + z_{12}(\mathbf{p}^*) - 2z_{13}(\mathbf{p}^*)\right).
\end{aligned}$$

and the corresponding eigenvectors ν_1 and ν_2 lie perpendicular to l_3 and in l_3 respectively. If an eigenvalue goes through -1 a period doubling bifurcation occurs in the dynamical system restricted to the center manifold (see Guckenheimer and Holmes (1983), p.158). The center manifold is locally $\mathbf{Z}_2$-symmetric (Kuznetsov (1995), Theorem 7.6).

[10] These two types of period doubling bifurcations correspond to pitchfork bifurcations of f^2. One of these pitchfork bifurcations is symmetry preserving ($\mu_2 = -1$) and the other one is symmetry breaking ($\mu_1 = -1$).

When $\mu_1 = -1$ the center manifold lies tangent to ν_1 and therefore perpendicular to the fixed point subspace. When $\mu_2 = -1$ the center manifold coincides with the fixed point subspace. We have $\mu_1 < \mu_2$ if $z_{12}(\mathbf{p}^*) > z_{13}(\mathbf{p}^*)$ and $\mu_1 > \mu_2$ if $z_{12}(\mathbf{p}^*) < z_{13}(\mathbf{p}^*)$. ■

According to proposition 2.4 there are two different types of period two orbits with $\mathbf{Z}_2$ symmetry. Examples of these two period two orbits are shown in Figures 2.4.b) and 2.4.c). Since the three prices have to sum up to 3 in every period an increase of one price has to be accompanied by the decrease of another price. Consider the period two orbit $\{\mathbf{p}^1, \mathbf{p}^2\}$ in Figure 2.4.b) that lies off l_3. This period two orbit has the property that the market for the third good is in equilibrium in every period but the first two prices fluctuate between the same values: when p_1 is high, p_2 is low and vice versa. In this case p_2 has to adjust to the fluctuations in p_1. This happens when $z_{12}(\mathbf{p}^*) > z_{13}(\mathbf{p}^*)$.

The period two orbit in Figure 2.4.c) that lies in l_3 has the property that prices of the first two goods are equal to each other in every period: in one period they are both high (above their equilibrium value) and in the next period they are both low (below their equilibrium value). In this case p_3 has to adjust to neutralize the fluctuations in p_1 (and p_2). This happens when $z_{13}(\mathbf{p}^*) > z_{12}(\mathbf{p}^*)$. So the burden of a change in p_1 is *completely* carried by the price of that good that is the closest substitute to good 1 (at the equilibrium price vector).

2.5.4 $\mathbf{D}_3$ symmetry and the Equivariant Branching Lemma

In the previous sections we saw that a rotational symmetry leads to a Hopf bifurcation and that a reflectional symmetry leads to a period doubling bifurcation of the equilibrium price vector when it loses stability. We now want to study the case where the adjustment process has both a rotational and a reflectional symmetry. We can then apply the Equivariant Branching Lemma for period doubling (Chossat and Golubitsky (1988)) to arrive at the following result.

Proposition 2.5 *Assume that the symmetric equilibrium price vector* $\mathbf{p}^* = (1,1,1)$ *of the tâtonnement process with rotational and reflectional symmetry group* $\Gamma_f = \mathbf{D}_3 = \{I, M_{312}, M_{213}\}$ *is stable for* $\lambda > 0$ *small enough.* $\mathbf{p}^*$ *undergoes a period doubling bifurcation as* λ *goes through*

$$\lambda^{bif} = -\frac{4}{3z_{11}(\mathbf{p}^*)}. \tag{2.21}$$

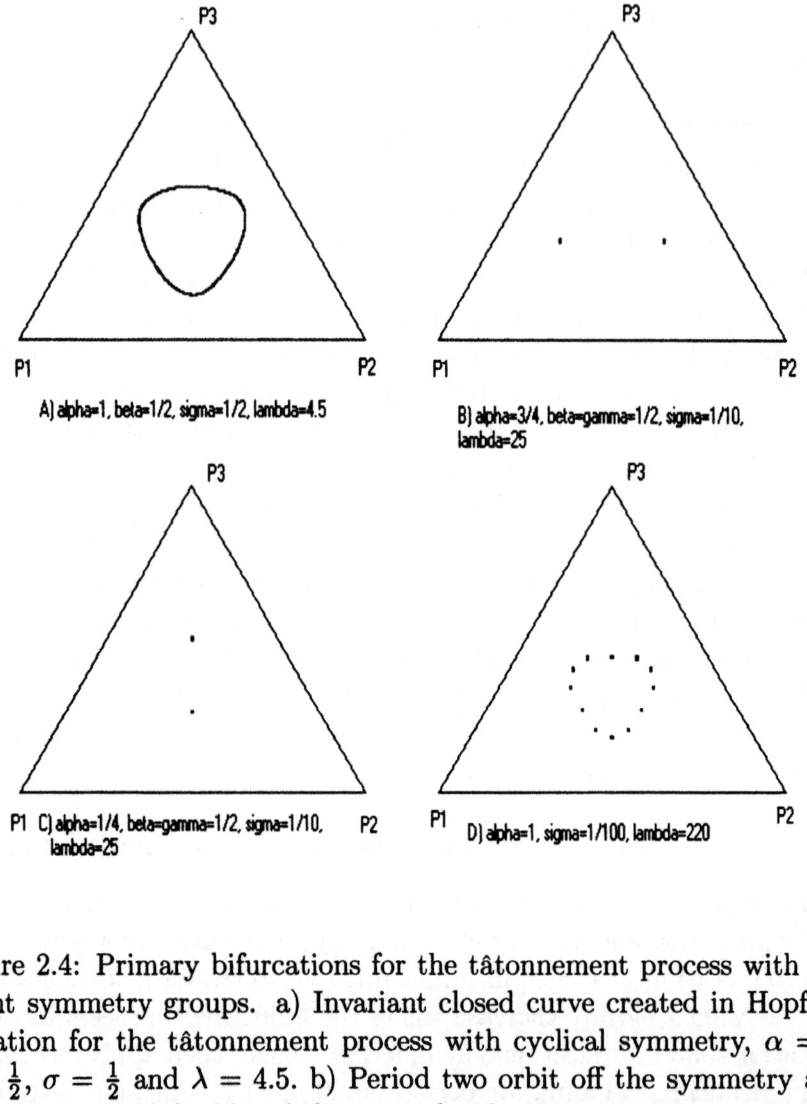

Figure 2.4: Primary bifurcations for the tâtonnement process with different symmetry groups. a) Invariant closed curve created in Hopf bifurcation for the tâtonnement process with cyclical symmetry, $\alpha = 1$, $\beta = \frac{1}{2}$, $\sigma = \frac{1}{2}$ and $\lambda = 4.5$. b) Period two orbit off the symmetry axis created in period doubling bifurcation for the tâtonnement process with reflectional symmetry, $\alpha = \frac{3}{4}$, $\beta = \gamma = \frac{1}{2}$, $\sigma = \frac{1}{10}$ and $\lambda = 25$. c) Period two orbit on symmetry axis created in period doubling bifurcation for the tâtonnement process with reflectional symmetry, $\alpha = \frac{1}{4}$, $\beta = \gamma = \frac{1}{2}$, $\sigma = \frac{1}{10}$ and $\lambda = 25$. d) Six period two orbits created in symmetry breaking period doubling bifurcation in tâtonnement process with cyclical and reflectional symmetry, $\alpha = 1$, $\sigma = \frac{1}{100}$ and $\lambda = 220$.

At this value of λ six period two orbits emerge, three stable and three unstable. For every fixed point subspace $l_i, i = 1, 2, 3$ there is a period two orbit on l_i and a period two orbit off l_i but with a reflectional symmetry with respect to l_i.

Proof. To proof Proposition 2.5 we use the following result (Chossat and Golubitsky (1988, Theorem 4.1)).

Theorem 2.1 Equivariant Branching Lemma for period doubling. *Consider a dynamical system $h : \mathbb{R}^n \to \mathbb{R}^n$ with a symmetry group Γ_h. Assume that Γ_h acts absolutely irreducibly on $\mathbb{R}^n$, that $\mathbf{J}_h^* = -I_n$ and that* $\dim Fix\,(M) = 1$, *for a subgroup $M \subset \Gamma_h \oplus \{\pm I_n\}$. Then generically, there exists a branch of period two points for h bifurcating at the fixed point and tangent to $Fix\,(M)$ at the fixed point. If $M \subset \Gamma_h$, then the branch lies in $Fix\,(M)$.*

Sketch of proof of Theorem 2.1. The symmetry group Γ_h is said to be acting *absolutely irreducible* on $\mathbb{R}^n$ if the only matrices commuting with all symmetries of Γ_h are multiples of the identity matrix. Then at a bifurcation all eigenvalues are equal to $+1$ or all eigenvalues are equal to -1. For a good understanding of the bifurcation that occurs when all eigenvalues are equal to -1, it is useful to discuss the case where all eigenvalues are $+1$ first. Recall that the fixed point subspace of a symmetry is an invariant space of the dynamical system. Now if $h\,(.)$ has a symmetry with a one dimensional fixed point subspace then the search for fixed points can be confined to this fixed point subspace. This gives the *Equivariant Branching Lemma* which states that for every subgroup of Γ_h that has a one dimensional fixed point subspace there exists a branch of fixed points of h bifurcating at the origin. This branch lies in this fixed point subspace (for a discussion of this theorem see Golubitsky, Stewart and Schaeffer (1988, Chapter XIII)).

If all eigenvalues are equal to -1 the analysis is more complicated. The normal form does not only commute with the symmetry group Γ_h but also with the linear part of the dynamical system which is $-I_n$ (Chossat and Golubitsky (1988)). Therefore the normal form commutes with $\Gamma_h \oplus \{\pm I_n\}$. Now we can apply the same reasoning as above, only we have to distinguish between subgroups with a one dimensional fixed point subspace that are in Γ_h and ones that are in $\Gamma_h \oplus \{\pm I_n\}$ but not in Γ_h. Therefore, there exists a branch of period two points bifurcating at the fixed point and tangent to $Fix\,(M)$, for each $M \in \Gamma_h \oplus \{\pm I_n\}$ with $\dim Fix\,(M) = 1$.

Proof of proposition 2.5. We can now use Theorem 2.1 to proof proposition 2.5. From proposition 2.2 it follows that the symmetry group $\mathbf{D}_3$ acts absolutely irreducible on $\mathbb{R}^3_+$. Furthermore $\mathbf{D}_3 \oplus \{\pm I_3\}$ contains six subgroups that have a one dimensional fixed point subspace. Three of these are also in $\mathbf{D}_3$, namely the three $\mathbf{Z}_2$ symmetries with fixed point subspaces l_1, l_2 and l_3, respectively. The three other fixed point subspaces l_1^*, l_2^* and l_3^* can be obtained by rotating l_1, l_2 and l_3 over $\frac{1}{2}\pi$. Application of Theorem 2.1 then gives the result. ■

For $\lambda > \lambda^{bif}$ six period two orbits exist, three of which are stable. Figure 2.4.d) shows the six period two orbits for a tâtonnement process with $\mathbf{D}_3$ symmetry ($\alpha = 1, \sigma = \frac{1}{100}, \lambda = 220$). In this case the three period two orbits off the symmetry axes are stable. Up till now we have seen that when the symmetric equilibrium price vector loses stability the resulting attractor has the same symmetry group as the symmetric equilibrium price vector. For the period two orbits emerging in the case where the tâtonnement process has both symmetries this no longer holds. These period two orbits only have a reflection symmetry. In fact rotating each of these period two orbits over $\frac{2}{3}\pi$ gives one of the other period two orbits. Thus the symmetry of the three coexisting period two orbits is smaller than the symmetry of the equilibrium price vector. Therefore the bifurcation described in proposition 2.5 is called a *symmetry-breaking* bifurcation.

2.6 Global dynamics

2.6.1 Symmetry-breaking and -increasing bifurcations

In Section 2.5.4 we have seen that when the symmetric equilibrium price vector of a tâtonnement process with both a reflectional and a rotational symmetry loses stability three attracting and coexisting period two orbits emerge. Each of these new attractors only has a $\mathbf{Z}_2$ symmetry. The symmetry group of the attractor of the tâtonnement process decreases from $\{I, M_{213}, M_{312}\}$ to either $\{I, M_{213}\}, \{I, M_{321}\}$ or $\{I, M_{132}\}$. However, the attractors in Figure 2.3 which come from tâtonnement processes with $\mathbf{D}_3$ symmetry are fully symmetric. So as λ increases at a certain point the symmetry group of the attractor has to increase again. This happens at a *symmetry increasing* bifurcation. Such bifurcations seem to be generic in tâtonnement processes with $\mathbf{D}_3$ symmetry.

Consider one of the stable period two orbits that emerges after the

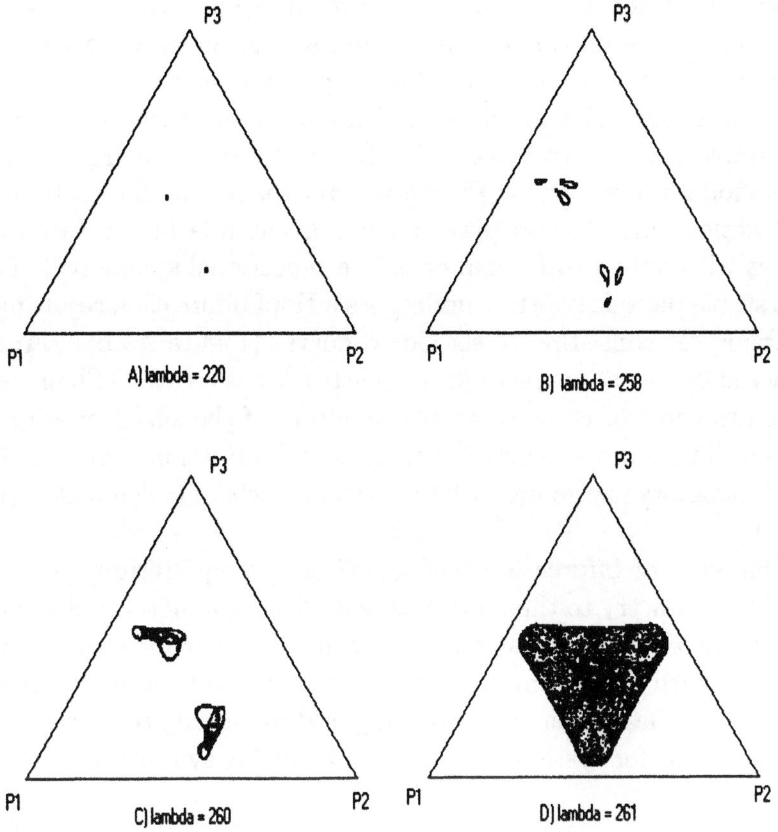

Figure 2.5: Bifurcation scenario for tâtonnement process with D_3 symmetry, $\alpha = 1$, $\sigma = \frac{1}{100}$. a) $\lambda = 220$, b) $\lambda = 258$, c) $\lambda = 260$, d) $\lambda = 261$.

symmetry breaking period doubling bifurcation. This period two orbit undergoes a bifurcation route to chaos, resulting in a $\mathbf{Z}_2$-symmetric strange attractor. By symmetry there are two other, conjugate $\mathbf{Z}_2$-symmetric strange attractors. As the speed of adjustment increases these attractors grow in magnitude. At a certain point these attractors with $\mathbf{Z}_2$ symmetry merge into one attractor which inherits all symmetries of the conjugate attractors and therefore has $\mathbf{D}_3$ symmetry again.

Figure 2.5 illustrates one such a bifurcation scenario.[11] After the symmetry breaking period doubling bifurcation there are three stable

[11]This particular bifurcation scenario corresponds to scenario 3 in Chossat and Golubitsky (1988).

period two orbits (one of them is shown in Figure 2.5.a)). The period two orbit undergoes a Hopf bifurcation and an invariant set consisting of two closed curves emerges. More important for the global dynamics is the emergence of a stable period six orbit in the neighbourhood of each stable period two orbit. This period six orbit emerges through a saddle-node bifurcation of f^6 (we will return to this bifurcation in the next section since it also plays an important role in the tâtonnement process with only a rotational or only a reflectional symmetry). Each of these stable period six orbits undergoes a Hopf bifurcation resulting in an attracting set consisting of six closed curves (Figure 2.5.b)). From this six closed curves a two piece strange attractor is created (Figure 2.5.c)). There are three of these strange attractors. If the speed of adjustment grows a little more a symmetry increasing bifurcation occurs and these three attractors merge into a large attractor with $\mathbf{D}_3$ symmetry (Figure 2.5.d)).

This kind of bifurcation scenario (from an equilibrium price vector with $\mathbf{D}_3$ symmetry to three attractors with $\mathbf{Z}_2$ symmetry and then again to an attractor with $\mathbf{D}_3$ symmetry) seems to occur for all tâtonnement processes with $\mathbf{D}_3$ symmetry. However as we will see in the next two subsections, these symmetry breaking and increasing bifurcation routes can also occur for systems with only a $\mathbf{Z}_2$ or $\mathbf{R}_3$ symmetry.

2.6.2 Rotational symmetry

From Section 2.5.2 we know that the economy with a rotational symmetry undergoes a Hopf bifurcation as the speed of adjustment increases. For $\lambda < \lambda^{bif}$ the equilibrium price vector $\mathbf{p}^* = (1,1,1)$ is locally stable. For $\lambda > \lambda^{bif}$ the equilibrium price vector is unstable and there is an attracting invariant closed curve. The dynamics on this curve can be periodic or quasi-periodic. Generically as λ increases the system undergoes numerous bifurcations resulting in periodic orbits on the closed curve. Then two periodic orbits, of the same period, arise on the closed curve, one stable and one unstable. The stable periodic orbit attracts almost all points on the closed curve and the dynamics are then periodic (this periodic orbit may however have a very long period). If for a certain value of the speed of adjustment such a periodic orbit does not exist the dynamics on the closed curve are *quasi-periodic*: the orbit of each point on the closed curve comes back arbitrarily close to that point but not exactly. The orbit fills up the closed curve completely. The time series of such an orbit is almost but not exactly periodic.

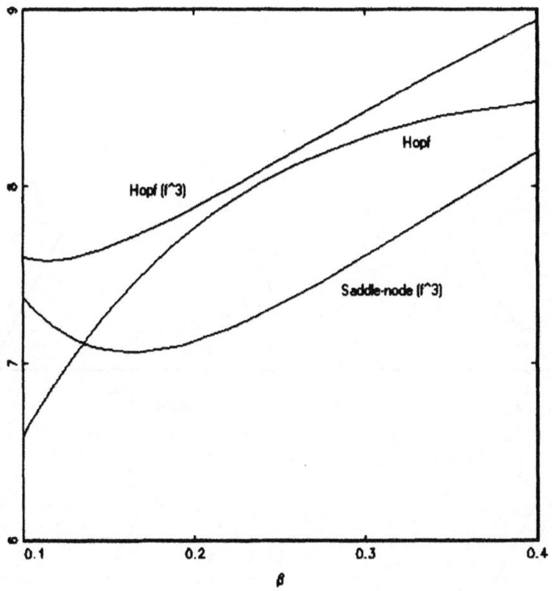

Figure 2.6: Bifurcation curves for tâtonnement process with cyclical symmetry, $\alpha = 1$ and $\sigma = \frac{1}{4}$.

For some regions of parameter values a period three orbit plays an important role. Figure 2.6 presents bifurcation curves for the model with $\alpha = 1$ and $\sigma = \frac{1}{4}$. These bifurcation curves show those combinations of β and λ at which a specific bifurcation occurs.

The curve denoted **Hopf** represents the Hopf-bifurcation of the symmetric equilibrium $\mathbf{p}^*$ as described above. The curve denoted **Saddle-node** (f^3) gives the values of β and λ at which a saddle-node bifurcation of the third iterate of the tâtonnement process takes place. At these values of β and λ and the prices $\mathbf{p}$ at which such a bifurcation occurs we have that one of the eigenvalues of the Jacobian of the system $\mathbf{p}_{t+3} = f^3(\mathbf{p}_t)$ becomes $+1$ and two period three orbits emerge, one stable and the other a saddle. Both period three orbits have symmetry M_{312}, that is, if this three cycle is $\{\mathbf{p}^1, \mathbf{p}^2, \mathbf{p}^3\}$, then $\mathbf{p}^2 = M_{312}\mathbf{p}^1$ and $\mathbf{p}^3 = M_{312}\mathbf{p}^2$. The curve **Hopf** ($f^3$) gives the values of β and λ for which the stable period three cycle undergoes a Hopf bifurcation and an invariant set consisting of three closed curves emerges. Now take for example $\beta = \frac{1}{4}$. Then from Figure 2.6 there are values λ^{sn3}, λ^h and λ^{h3} of λ with $\lambda^{sn3} < \lambda^h < \lambda^{h3}$ such that we have the following: for $\lambda < \lambda^{sn3}$

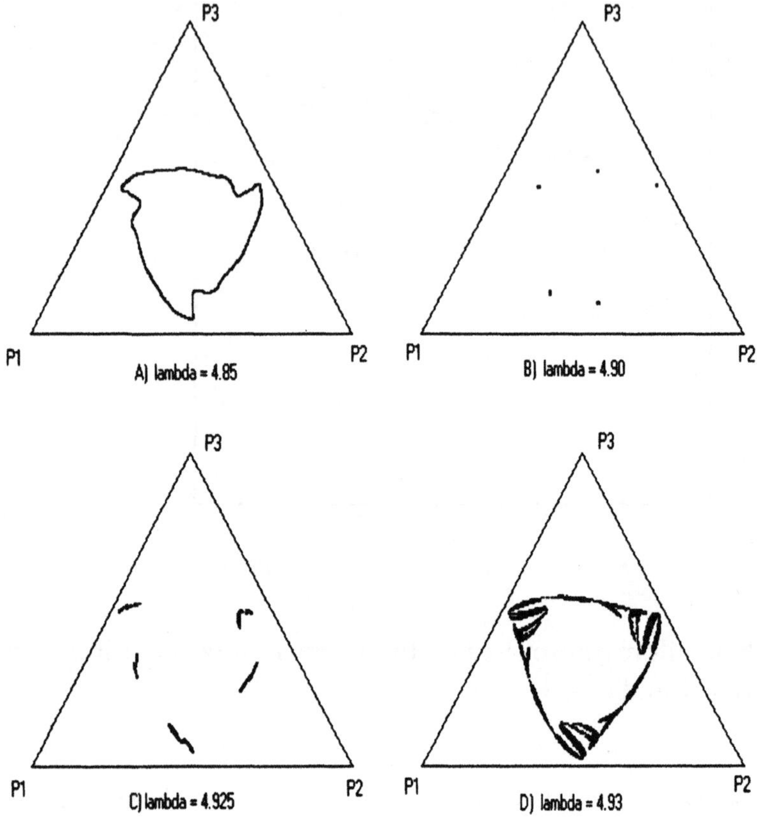

Figure 2.7: Bifurcation scenario for tâtonnement process with cyclical symmetry, $\alpha = 1$, $\beta = \frac{1}{2}$, $\sigma = \frac{1}{2}$. a) $\lambda = 4.85$, b) $\lambda = 4.90$, c) $\lambda = 4.925$ and d) $\lambda = 4.93$.

the symmetric equilibrium is the unique attractor: for all initial price vectors the tâtonnement process converges to $\mathbf{p}^*$. For $\lambda = \lambda^h$ the symmetric equilibrium price vector becomes unstable. For $\lambda^{sn3} < \lambda < \lambda^h$ there are two attractors: the symmetric equilibrium and a stable period three orbit. For $\lambda^h < \lambda < \lambda^h + \varepsilon < \lambda^{h3}$ there are also two attractors: an invariant closed curve around the unstable symmetric equilibrium and a stable period three orbit. At a certain value of λ $\left(\lambda^h < \lambda < \lambda^{h3}\right)$ the invariant closed curve loses stability and all points converge to the period three orbit. For $\lambda > \lambda^{h3}$ there is an attracting invariant set consisting of three closed curves. As λ increases even more this three piece attracting invariant set bifurcates into a three-piece strange attractor (an example of an attractor created in this way is the one shown in Figure 2.1.b)). The behaviour of the tâtonnement process for large λ is governed by what happens with the period three orbit instead of what happens with the symmetric equilibrium. Similar bifurcation scenarios, only then involving a period six orbit emerging in the neighbourhood of a period two orbit, occur for the tâtonnement process with $\mathbf{D}_3$ or $\mathbf{Z}_2$ symmetry.

We now discuss a bifurcation scenario for one particular tâtonnement process with cyclical symmetry. We take $\alpha = 1, \beta = \frac{1}{2}$ and $\sigma = \frac{1}{2}$. In this case a Hopf bifurcation occurs at $\lambda^{bif} = \frac{35+21\sqrt{2}}{11+3\sqrt{2}} \approx 4.2446$. For values of λ larger than λ^{bif} an attracting closed curve exists. As λ increases this closed curve becomes larger. For a certain value of the speed of adjustment ($\lambda \approx 4.8955$) a saddle node bifurcation of f^5 on the curve occurs and two period 5 orbits emerge, one stable and one unstable. This period 5 cannot have a cyclical symmetry, so there have to be two other stable period 5 and two other unstable period 5 orbits. Since these orbits have no nontrivial symmetry this is a symmetry breaking saddle node bifurcation resulting in three coexisting attractors. These coexisting period 5 orbits undergo a cascade of period doubling bifurcations resulting in three coexisting five piece strange attractor. As the speed of adjustment increases even more ($\lambda \approx 4.9278$) these three coexisting strange attractors merge into one large attractor. This attractor has a cyclic symmetry again and the last bifurcation therefore is a symmetry increasing bifurcation. Figure 2.7 shows the attractors for some values of λ for this specific example.

2.6.3 Reflection symmetry

We now want to see what happens to the period two orbits in the tâtonnement process with $\mathbf{Z}_2$ symmetry when the speed of adjustment λ increases. We distinguish between the two different period two orbits that were discussed in Section 2.5.3.

First consider the period two orbit that lies in the fixed point subspace l_3. This period two orbits consists of two symmetric equilibria of f^2. Since f^2 has the same symmetry group as f it follows that the eigenvalues of the Jacobian of f^2 evaluated at one of the fixed points are real and a Hopf bifurcation cannot occur. But in this case we cannot ignore the possibility of an eigenvalue going through $+1$ as λ increases. There seem to be two possible bifurcations: a period doubling bifurcation resulting in a stable period four orbit, or a pitchfork bifurcation resulting in three period two orbits , two of which are stable. Let $\{\mathbf{p}^1, \mathbf{p}^2\} \in l_3$ be the period two orbit. It is easy to see that the eigenvalues η_1 and η_2 of the Jacobian of f^2 evaluated at $\mathbf{p}^1$ are equal to the product of the eigenvalues of the Jacobians of f evaluated at $\mathbf{p}^1$ and $\mathbf{p}^2$: $\eta_1 = \mu_1(\mathbf{p}^1)\,\mu_1(\mathbf{p}^2)$ and $\eta_2 = \mu_2(\mathbf{p}^1)\,\mu_2(\mathbf{p}^2)$. The eigenvector corresponding to η_1 lies perpendicular to the fixed point subspace and the eigenvector corresponding to η_2 lies in the fixed point subspace. It is important again which eigenvalue goes through ± 1 first. Numerically we only observe the following two bifurcations as λ increases:

- $\eta_1 = 1$: a symmetry breaking pitchfork bifurcation of f^2 occurs. The symmetric period two orbit loses stability and the two new period two orbits are stable and asymmetric,
- $\eta_2 = -1$: a period doubling bifurcation of f^2 occurs. The period two orbit loses stability and a stable period four orbit emerges. This new attractor lies in the fixed point subspace.

Of these two the latter occurs more often. This period doubling bifurcation is followed by a cascade of period doubling bifurcations leading into a one-dimensional attractor that fills up part of the fixed point subspace. An example of such an attractor is given in Figure 2.2.a).

Now consider the case where the stable period two orbit created through the period doubling bifurcation of the equilibrium price vector does not lie in the fixed point subspace. In this case $\mathbf{p}^1$ and $\mathbf{p}^2$ correspond to asymmetric fixed points of f^2. Nothing much can be said about this. However, from simulations we found two important secondary bifurcations, reminiscent of the bifurcation in the tâtonnement process with rotational symmetry. One is the Hopf bifurcation and the emergence of an attractor consisting of two invariant cycles. The other possibility is this same Hopf bifurcation accompanied by a saddle-node bifurcation of f^6 in the neighbourhood of the period two orbit resulting in an attracting period six orbit. The behaviour of this period six orbit then governs what happens to the tâtonnement process as the speed of adjustment grows even more. An example of an attractor created from such a process is shown in Figure 2.2.b).

Figure 2.8 shows bifurcation curves for the tâtonnement process with symmetry group $\{I, M_{213}\}$ and $\beta = \gamma = \frac{1}{2}, \sigma = \frac{1}{10}$. Curve 1 corresponds to combinations of α and λ at which μ_1 equals -1 and curve 2 corresponds to combinations of α and λ at which μ_2 equals -1. The two

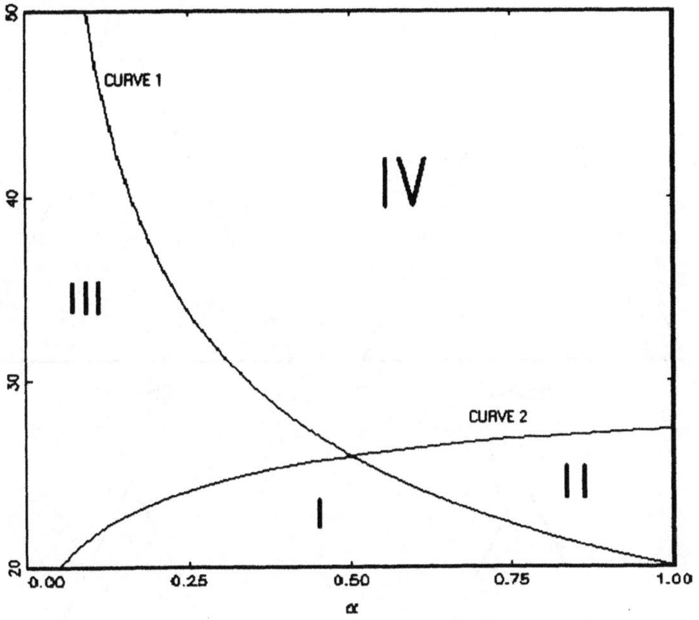

Figure 2.8: Period doubling bifurcation curves for the tâtonnement process with reflectional symmetry, $\beta = \frac{1}{2}$, $\gamma = \frac{1}{2}$ and $\sigma = \frac{1}{10}$.

curves intersect at the point a where $\alpha = \frac{1}{2}$. We obtain the following information from this diagram. The equilibrium price vector $\mathbf{p}^*$ is locally stable if (α, λ) lies in area **I**. If (α, λ) lies in **II** or **III** then $\mathbf{p}^*$ is a saddle point. In (the lower parts of) **II** there is a stable period two orbit off of l_3 and in (the lower parts of) area **III** there is a stable period two orbit on l_3. In area **IV** the equilibrium price vector $\mathbf{p}^*$ is a repellor. With respect to the second bifurcation we have the following numerical results: for $\alpha \in (0, \alpha_0)$ with $\alpha_0 \approx 0.057$ the period two orbit on the symmetry axis undergoes a pitchfork bifurcation resulting in three period two orbits, one unstable and on the symmetry axis and two stable and off of the symmetry axis. For $\alpha \in \left(\alpha_0, \frac{1}{2}\right)$ the second bifurcation is a period doubling bifurcation resulting in a stable period four orbit in the fixed point subspace l_3. For $\alpha \in \left(\frac{1}{2}, 1\right]$ the stable period two orbit lies off of the symmetry axis and f^2 undergoes a Hopf bifurcation.

At the point a ($\alpha = \frac{1}{2}$) both eigenvalues are equal to -1: $\mathbf{p}^*$ goes immediately from being an attractor to being a repellor. Notice that, since $\beta = \gamma = \frac{1}{2}$, for $\alpha = \frac{1}{2}$ we have a tâtonnement process with $\mathbf{D}_3$ symmetry. However the case where both eigenvalues go through -1 simultaneously

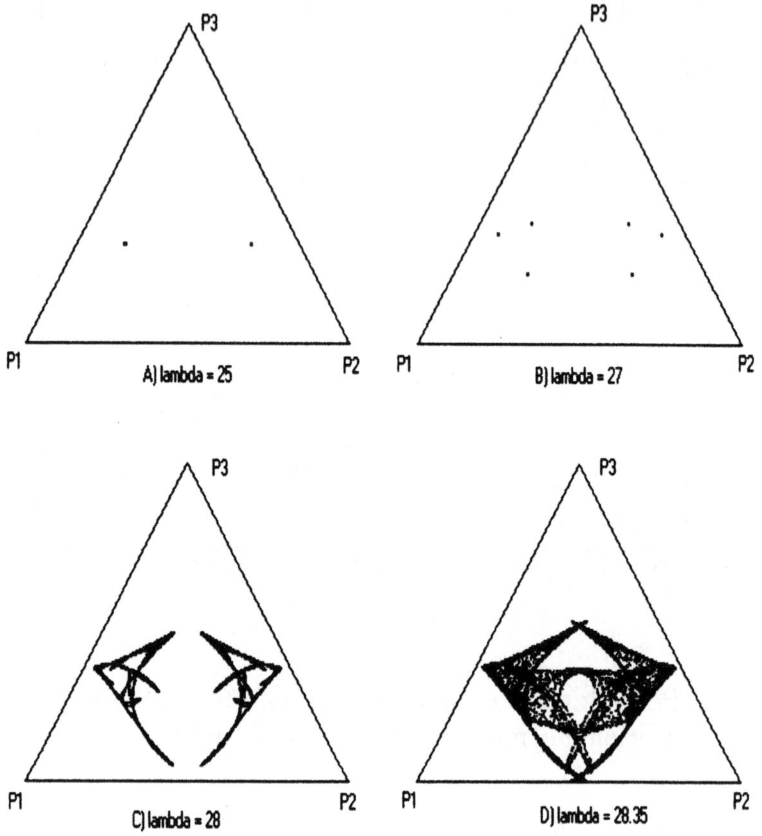

Figure 2.9: Bifurcation scenario for tâtonnement process with reflectional symmetry $\alpha = 1$, $\beta = \frac{1}{2}$, $\gamma = \frac{1}{2}$, $\sigma = \frac{1}{10}$. a) $\lambda = 25$, b) $\lambda = 27$, c) $\lambda = 28$ and d) $\lambda = 28.35$.

occurs also in "proper" $\mathbf{Z}_2$-symmetric tâtonnement processes when $\beta \neq \gamma$ and $z_{12}(\mathbf{p}^*) = z_{13}(\mathbf{p}^*)$. Then we have a so called codimension 2 bifurcation in the space of $\mathbf{Z}_2$-symmetric tâtonnement processes.

Now we consider one particular bifurcation scenario. Figure 2.9 shows the bifurcation scenario for the case with $\alpha = 1, \beta = \frac{1}{2}, \gamma = \frac{1}{2}$ and $\sigma = \frac{1}{10}$. For $\lambda^{bif} \approx 20.01$ a period doubling bifurcation occurs resulting in a period two orbit off l_3. As λ increases two bifurcations occur: the period two orbit undergoes a Hopf bifurcation (at $\lambda \approx 26.23$) resulting in an attracting invariant set consisting of two closed curves, and a saddle node bifurcation of f^6 occurs (this happens at $\lambda \approx 26.88$) resulting in

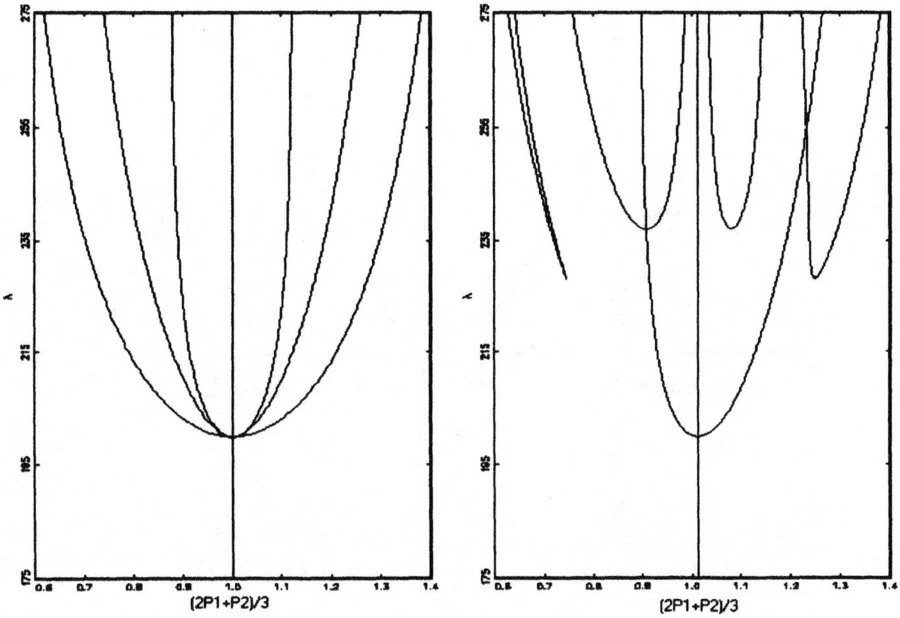

Figure 2.10: a) Stable period two curves for tâtonnement process with D_3 symmetry $\left(\alpha_{ij} = 1, \sigma = \frac{1}{100}\right)$ created through symmetry breaking period doubling bifurcation. b) Period two curves in a tâtonnement process close to D_3 symmetry ($\alpha_{ij} = 1$, $\alpha_{12} = 0.95$, $\alpha_{13} = 0.9$, $\sigma = \frac{1}{100}$).

two period 6 orbits, one stable and one unstable. The stable period 6 orbit undergoes a cascade of period doubling bifurcations resulting in a two piece strange attractor (Figure 2.9.c)). As the speed of adjustment λ increases even more this two piece strange attractor becomes a one piece strange attractor (Figure 2.9.d)). Notice that if we consider the dynamical system f^2 then the $\mathbf{Z}_2$ symmetric two piece strange attractor of f corresponds to two asymmetric, conjugate attractors of f^2 and the change of this two piece attractor of f^2 into a one piece attractor corresponds to a symmetry increasing bifurcation.

2.7 An asymmetric price adjustment process

Thus far we concentrated on global dynamics of symmetric tâtonnement processes. Depending on this symmetry, different kinds of bifurcations occur. In particular a bifurcation from a stable steady state to six coexisting period two orbits occurs in the tâtonnement process with $\mathbf{D}_3$

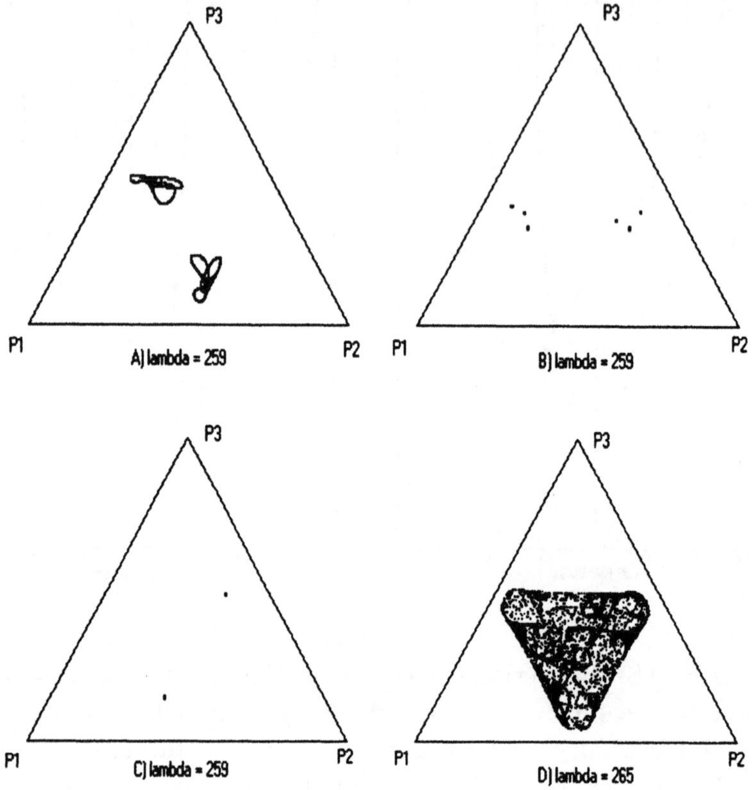

Figure 2.11: Attractors for asymmetric tâtonnement process. a)-c) three different coexisting attractors for $\lambda = 259$. d) unique attractor for $\lambda = 265$.

symmetry. Then, as the equilibrium price vector loses stability, three coexisting attracting period two orbits emerge. Obviously, a price adjustment process with a $\mathbf{D}_3$ symmetry is very special. The question thus arises whether the analysis of a special symmetric tâtonnement process is useful in order to understand the global dynamics of general asymmetric tâtonnement processes. In this section we show that nearby asymmetric tâtonnement processes exhibit bifurcation routes to strange attractors very similar to the special symmetric bifurcation routes. Therefore, symmetric tâtonnement processes can serve as an "organizing centre" in the analysis of the global dynamics in general tâtonnement processes.

We study a numerical example with the following parameter values

$$A = \begin{pmatrix} 1 & 0.95 & 0.9 \\ 1 & 1 & 1 \\ 1 & 1 & 1 \end{pmatrix}, \sigma = \frac{1}{100}, \tag{2.22}$$

and initial endowments as before.

Notice that the multiplicative price adjustment process with these preferences has no nontrivial symmetry, but that it is close to a price adjustment process with $\mathbf{D}_3$ symmetry.

The equilibrium price vector is $\mathbf{p}^* = (1.0177656, 1.0002052, 0.9820292)$ and the eigenvalues of the Jacobian evaluated at this equilibrium price vector are $\mu_1 = 1 - 0.00982501\lambda$ and $\mu_2 = 1 - \frac{1}{100}\lambda$, respectively. A period doubling bifurcation occurs at $\lambda^{bif} = 200$ resulting in a stable period two orbit. Only one period two orbit emerges. However as λ increases to $\lambda^{sn1} \approx 228.2$ the second iterate of the price adjustment process, f^2, undergoes a saddle node bifurcation and two period two orbits emerge, one stable and one unstable. As λ increases even more to $\lambda^{sn2} \approx 237.0$ f^2 undergoes a second saddle-node bifurcation again resulting in a stable and an unstable period two orbit. Just as in the price adjustment process with $\mathbf{D}_3$ symmetry three coexisting stable period two orbits exist. In this case however, these period two orbits are not created through a single bifurcation at a single value of the bifurcation parameter λ, but through a series of bifurcations at different values of λ. Figure 2.10 shows two bifurcation diagrams. Figure 2.10.a) shows the three stable period two orbits created in the symmetry breaking period doubling bifurcation for the tâtonnement process with $\mathbf{D}_3$ symmetry ($\alpha_{ij} = 1, \sigma = \frac{1}{100}$) and Figure 2.10.b) shows three stable period two orbits and three unstable period two orbits created in the period doubling and saddle-node bifurcations for the asymmetric tâtonnement process described above.

Each of the three coexisting stable period two orbits undergoes a bifurcation route to a strange attractor, with bifurcations occurring for different values of the speed of adjustment λ. Therefore for certain values of λ there are three different coexisting attractors. Figure 2.11.a)-c) shows the three different coexisting attractors for $\lambda = 259$. As λ increases more the three coexisting attractors merge, just as in the symmetric case, to one large attractor. This attractor is shown in Figure 2.11.d) for $\lambda = 263$.

This example illustrates that the bifurcation scenario of the price adjustment process with a $\mathbf{D}_3$ symmetry might tell us something about the

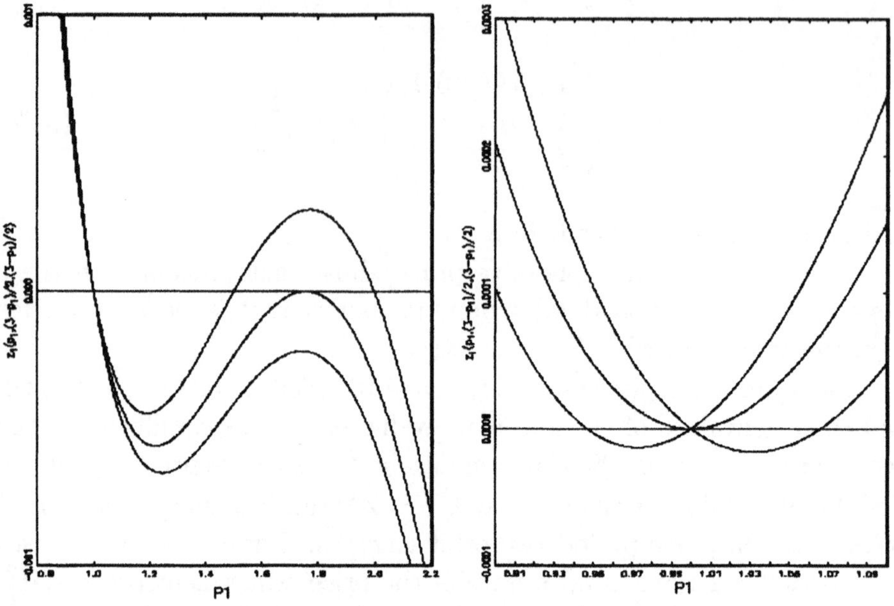

Figure 2.12: Aggregate demand of good 1 on l_1. a) Saddle-node bifurcation. b) Transcritical bifurcation.

behaviour of a price adjustment process that has no nontrivial symmetry but that is close to a price adjustment process with $\mathbf{D}_3$ symmetry.

2.8 Multiplicity of equilibria

2.8.1 Transcritical and saddle-node bifurcations: a case study

Up till now we have focused attention on what happens with the equilibrium price vector if the speed of adjustment is increased. We have seen that an attracting closed curve, a stable period two orbit or three coexisting stable period two orbits emerge, depending on the symmetry of the economy, when the equilibrium price vector loses its stability. These bifurcations take place when the eigenvalues are complex conjugates and $|\mu_{1,2}| = 1$, when they are real and $\min\{\mu_1, \mu_2\} = -1$ or when they are real and $\mu_1 = \mu_2 = -1$, respectively. We also argued that bifurcations where one of the eigenvalues goes through $+1$ do not occur when the speed of adjustment λ increases since λ cannot influence the number of equilibria. However, an eigenvalue can equal $+1$ as another parameter varies. In this section we will show that a saddle-node and a transcritical bifurcation occurs in our tâtonnement process with symmetry group $\Gamma_f = \mathbf{D}_3$ as α decreases. Because we cannot find an explicit expression for the asymmetric equilibrium price vectors in terms of α we present a numerical analysis. We will investigate the case with $\sigma = \frac{1}{10}$. Our analysis is somewhat similar to Bala (1997) who shows that for the continuous tâtonnement process in a symmetric exchange economy with two commodities and CES utility functions the symmetric equilibrium price vector undergoes a pitchfork bifurcation for certain parameter values.

The only fully symmetric equilibrium price vector is $\mathbf{p}^*=(1,1,1)$ which is an equilibrium for all values of α and σ. The new equilibria must therefore have a smaller symmetry group. We will restrict attention to the set $l_1 = \{\mathbf{p} \in \mathbf{S}_3 \mid p_2 = p_3\}$. Using symmetry we find that $z_2(\mathbf{p}) = z_3(\mathbf{p})$ for $\mathbf{p} \in l_1$ and from Walras' Law we find that $p_1 z_1(\mathbf{p}) = -2p_2 z_2(\mathbf{p})$ for $\mathbf{p} \in l_1$. Hence a zero of $z_1(\mathbf{p})$ with $\mathbf{p} \in l_1$ corresponds to an equilibrium price vector. Notice that such an equilibrium has symmetry M_{132} and that therefore there have to be two other asymmetric equilibria related to this one by symmetry. We now have the possibility to study the multiplicity of equilibria (in l_1) with the aid of simple diagrams (such a graphical analysis is always possible for a general equilibrium model with two commodities, see for example Bala (1997) and one of the examples in the next chapter). Figure 2.12 shows the function $\widehat{z}_1(p_1) \equiv z_1\left(p_1, \frac{3-p_1}{2}, \frac{3-p_1}{2}\right)$ for $\sigma = \frac{1}{10}$ and different values of α. Figure 2.12.a) illustrates a subcritical saddle-node bifurcation. At $\alpha^{sn} \approx 0.051138$ a new equilibrium $\mathbf{p}^{sn} \approx (1.7556, 0.6222, 0.6222)$ emerges. The eigenvalues at this price vector are $\mu_1 \approx 1$ and $\mu_2 \approx 1 - 0.034\lambda$. For $\alpha > \alpha^{sn}$ there are no equilibrium price vectors in the neighbourhood of $\mathbf{p}^{sn}$ and for $\alpha < \alpha^{sn}$ there are two equilibrium price vectors in the neighbourhood of $\mathbf{p}^{sn}$. The new equilibrium price vector with $\widehat{z}_{11}(p_1) < 0$ is locally stable for λ small enough, the new equilibrium price vector with $\widehat{z}_{11}(p_1) > 0$ is unstable for all $\lambda > 0$. Due to symmetry two other saddle-node bifurcations occur at α^{sn} and price vectors $\mathbf{p} = M_{312}\mathbf{p}^{sn}$ and $\mathbf{p} = M_{231}\mathbf{p}^{sn}$.

As α decreases further the distance between the new asymmetric equilibria in Figure 2.12.a) increases until, at a certain value of α, one of the unstable equilibrium price vectors collides with the symmetric equilibrium $\mathbf{p}^*$ and changes stability, as shown in Figure 2.12.b). At $\mathbf{p}^*$ we have $\mu_{1,2} = 1 + \frac{3}{2}\lambda \widehat{z}_{11}(\mathbf{p}^*)$. The transcritical bifurcation occurs when

$z_{11}(\mathbf{p}^*) = 0$. This happens at $\alpha^{tc} = \left(\frac{1-2\sigma}{1+\sigma}\right)^{\frac{1}{\sigma}} = \left(\frac{8}{11}\right)^{10} \approx 0.041397$. For $\alpha > \alpha^{tc}$ the symmetric equilibrium price vector $\mathbf{p}^*$ is locally stable for small enough $\lambda > 0$. For $\alpha < \alpha^{tc}$ the symmetric equilibrium price vector is unstable and the asymmetric equilibrium price vector is stable. Because of symmetry, at $\alpha = \alpha^{tc}$ four equilibria collide and change stability. Notice that this is a straightforward application of the Equivariant Branching Lemma for fixed points.

We can summarize the bifurcation scenario for an economy with $\Gamma_f = \mathbf{D}_3$ and $\sigma = \frac{1}{10}$ in the following table (keep in mind that this table only contains information about the equilibria that lie on the symmetry axes).

α	# equilibria (# locally stable equilibria)
$\alpha > \alpha^{sn}$	1(1)
$\alpha = \alpha^{sn}$	4(1)
$\alpha^{tc} < \alpha < \alpha^{sn}$	7(4)
$\alpha = \alpha^{tc}$	4(3)
$0 < \alpha < \alpha^{tc}$	7(6)

Here stability has to be interpreted as follows: there is a $\widehat{\lambda} > 0$ such that for every $\lambda \in \left(0, \widehat{\lambda}\right)$ the equilibrium price vector is locally stable. Of course, as the speed of adjustment increases enough all equilibrium price vectors become unstable.

Notice that the bifurcations in this subsection only depend on the aggregate demand functions and not on the adjustment rule.

2.8.2 The continuous tâtonnement process

It may be useful to compare the results of the last subsection with the continuous tâtonnement process. For continuous dynamical systems an equilibrium is locally stable if the real parts of the eigenvalues are negative. A bifurcation of the continuous system occurs when an eigenvalue crosses the imaginary axis. If the eigenvalue is real for example a saddle-node or a transcritical bifurcation occurs and if the eigenvalues lie on the imaginary axis a Hopf bifurcation occurs. The eigenvalues $\eta_{1,2}$ of the continuous tâtonnement process $\frac{dp_i}{d_t} = z_i(\mathbf{p}(t))$ are related to the eigenvalues of our discrete tâtonnement process (2.7) in the following way: $\eta_i = \frac{\mu_i - 1}{\lambda}$, where λ is the speed of adjustment of the discrete tâtonnement process and μ_i is an eigenvalue of the discrete tâtonnement process. Hence the saddle-node and transcritical bifurcations described above also occur in the continuous tâtonnement process. For an economy with rotational symmetry, $\Gamma_f = \{I, M_{312}\}$, a Hopf-bifurcation occurs in the continuous tâtonnement process when $z_{11}(\mathbf{p}^*) = 0$, since in that case both eigenvalues lie on the imaginary axis.

These cases are similar to Scarf's well-known examples of global instability (Scarf (1960)). Scarf proves that for the continuous tâtonnement process with a $\mathbf{R}_3$ symmetry the equilibrium price vector is globally unstable if $z_{11}(\mathbf{p}^*) > 0$. In this case the (continuous or discrete) tâtonnement process (which is based on the assumption that aggregate demand for a good falls when its price increases) moves away from the equilibrium price vector. So the instability in the examples of Scarf and in the cases studied in the previous subsection arises from the form of the aggregate excess demand functions. In the case with "well-behaved" aggregate demand functions the continuous tâtonnement process is stable but the discrete tâtonnement process can be unstable due to overshooting: for λ too high the tâtonnement process jumps too far over the equilibrium price vector. Instability in this case results from the specification of the adjustment rule.

2.9 Summary and conclusions

Price normalization plays an important role in the dynamics of the tâtonnement process. The tâtonnement process can be stable under one normalization rule and unstable under another. Furthermore, by introducing a normalization rule some information is lost. We have argued that for the multiplicative price adjustment process, using the simplex as a normalization rule is natural since the multiplicative price adjustment process leaves the simplex invariant.

In studying the dynamical behaviour of our tâtonnement process we have focused on the cases where some nontrivial symmetry is present. We find that if the tâtonnement process has

- a rotational symmetry the symmetric equilibrium loses stability through a Hopf bifurcation,
- a reflectional symmetry the symmetric equilibrium loses stability through a period doubling bifurcation,
- both symmetries the symmetric equilibrium price vector loses stability through a symmetry breaking period doubling bifurcation.

In this last case the symmetry breaking bifurcation is followed by an infinite sequence of bifurcations leading to three coexisting conjugate strange attractors eventually resulting in a unique strange attractor again. Eventually a symmetry increasing bifurcation occurs where the three conjugate attractors merge into one unique attractor, which is fully symmetric again. We also encountered symmetry breaking and increasing bifurcations when the tâtonnement process has only a cyclical or reflectional symmetry. Obviously the symmetry of the tâtonnement process plays an important role in the dynamics of that tâtonnement process. Studying symmetric tâtonnement processes has been a first step, and may serve as an organizing centre in understanding the possible bifurcation routes in the multiplicative tâtonnement process on the simplex. The example in Section 2.7 shows that the bifurcation scenario in a symmetric tâtonnement process can tell us something about the bifurcation scenario in a tâtonnement process without symmetries.

The study of symmetric general equilibrium models has lead to another interesting result. It appears that the asymmetric equilibria of a general equilibrium model with a cyclical symmetry (like the ones that were found in Section 2.8) correspond to equilibrium cycles in a related overlapping generations model. This relationship is the subject of our next chapter.

Chapter 3

Perfect Foresight Cycles in Overlapping Generations Models

3.1 The overlapping generations model

Static general equilibrium models of the type discussed in the previous chapter are not particularly well suited for the analysis of dynamical issues. An alternative is the overlapping generations model, which has become a popular tool to study issues in monetary economics, public finance, business cycle theory and other fields in economics. The overlapping generations model is one of the first models in which it was shown that cyclical and chaotic behaviour can occur in equilibrium models with utility maximizing agents, market clearing in each period and agents that have perfect foresight. These perfect foresight cycles in overlapping generations models correspond to asymmetric equilibria of cyclical general equilibrium models of the type studied in the previous chapter. This equivalence is the subject of this chapter. But first, as a reminder, we will briefly discuss a simple example of an overlapping generations model with three generations.[1]

In each period t a new generation of agents that live for three periods is born. For simplicity we assume that all members of the generation are identical and that their number is constant. Agents of different gen-

[1] Although most of the applications of the overlapping generations model deal with two generations a number of properties only can be seen clearly in a model with at least three generations.

erations have the same preferences and endowments and there is only one commodity, which cannot be stored. In particular, an agent born in period t has initial endowment $\mathbf{w} = (w_0, w_1, w_2)$, where w_s is his endowment in period $t+s$, and a lifetime utility function $U\left(c_0^t, c_1^t, c_2^t\right)$, where c_s^t is consumption in period $t+s$. An agent born in period t faces prices p_t, p_{t+1} and p_{t+2}, when "young", "middle-aged" and "old" respectively and maximizes his lifetime utility $U\left(c_0^t, c_1^t, c_2^t\right)$ subject to his lifetime budget constraint

$$p_t c_0^t + p_{t+1} c_1^t + p_{t+2} c_2^t \leq p_t w_0 + p_{t+1} w_1 + p_{t+2} w_2.$$

This gives demand functions

$$d_0\left(p_t, p_{t+1}, p_{t+2}, \mathbf{w}\right), \ d_1\left(p_t, p_{t+1}, p_{t+2}, \mathbf{w}\right) \text{ and } d_2\left(p_t, p_{t+1}, p_{t+2}, \mathbf{w}\right).$$

Notice that agents of different generations have the same demand functions but face different prices during their lifetime. In each period t there are a young, a middle-aged and an old generation alive. Equilibrium on the market for the good in period t then requires

$$\begin{aligned} d_0\left(p_t, p_{t+1}, p_{t+2}, \mathbf{w}\right) + d_1\left(p_{t-1}, p_t, p_{t+1}, \mathbf{w}\right) + \\ d_2\left(p_{t-2}, p_{t-1}, p_t, \mathbf{w}\right) = w_0 + w_1 + w_2. \end{aligned} \tag{3.1}$$

An equilibrium path under perfect foresight is a price path that satisfies (3.1) for all t. Since only relative prices matter this equilibrium condition could also be formulated in terms of interest factors $\rho_t \equiv \frac{p_t}{p_{t+1}}$ as

$$d_0\left(\rho_t, \rho_{t+1}, \mathbf{w}\right) + d_1\left(\rho_{t-1}, \rho_t, \mathbf{w}\right) + d_2\left(\rho_{t-2}, \rho_{t-1}, \mathbf{w}\right) = w_0 + w_1 + w_2. \tag{3.2}$$

Given ρ_t, ρ_{t-1} and ρ_{t-2} (3.2) determines the interest factor ρ_{t+1}. We are dealing with a so-called *temporary equilibrium model*, where in each period a new price or interest factor emerges that clears the goods market in that period, given that past prices or interest factors are such that the markets cleared in previous periods. If the interest factor ρ_{t+1} determined by condition (3.2) exists and is unique (which is not always the case) it is implicitly given by the temporal equilibrium map

$$\rho_{t+1} = F\left(\rho_t, \rho_{t-1}, \rho_{t-2}\right). \tag{3.3}$$

Often it will be difficult or impossible to find an explicit expression for F. All paths generated by (3.2) or (3.3) are equilibrium paths. If almost all of them are feasible, that is, a solution exists for all t, then we have a continuum of equilibria. The overlapping generations model is then called *indeterminate* in the sense that its equilibria are not even locally unique (for a comprehensive discussion of determinateness of dynamic economic models see Farmer (1993)). In the literature the main focus has been on stationary equilibrium paths. It can be easily seen that there is always a solution with a constant price level p (and a corresponding interest factor $\rho^* = 1$). Condition (3.1) then becomes

$$d_0(p,p,p,\mathbf{w}) + d_1(p,p,p,\mathbf{w}) + d_2(p,p,p,\mathbf{w}) = w_0 + w_1 + w_2.$$

Multiplying both sides with p shows that this condition coincides with the budget constraint for an individual agent and therefore is always satisfied. Notice that $\rho^* = 1$ implies an interest rate of 0.[2] Now we have to address the following issue. Implicit in the construction of the model is the assumption that agents can trade commodities with each other. This is not the case. Take for example the situation where an agent receives all of his endowment when he is young and none when he is middle-aged or old. If he has convex preferences he wants to trade some of his endowment when he is young for consumption in the later periods of his life. The only agents he could trade with are the presently middle-aged and old. But they have nothing to offer him, so these trades cannot take place. The young agent wants some of the good in the future from the (unborn) young generations of the next two periods in order to give some of his endowment to the present middle-aged and old in this period. Therefore there has to be some institution (for example money or debt) that can realize trades between agents that are not able to trade directly with each other. The steady state discussed above therefore is called the *monetary* steady state. Aggregate savings in the monetary steady state can be positive or negative. Gale (1973) calls overlapping generations models where aggregate savings are positive in the monetary steady state *Samuelson* and overlapping generations models where they are negative *classical.* There are also steady states where aggregate savings are 0 and trade can be executed directly between agents. These steady states are called *balanced.* For the model with two generations it is clear that the only balanced steady state corresponds to

[2] If we would allow the population to grow at a constant rate β, then in this equilibrium prices would decrease at rate β^{-1} and the steady state interest factor would be β. The interest rate of an overlapping generations economy therefore only depends upon the biological growth rate and is not determined by economic factors, such as investment possibilities or discounting of the future by young consumers.

autarchy where each generation consumes its own endowment and no trade takes place.[3] For overlapping generations models with more than two generations (or several commodities per period and different agents per generation) balanced steady states do in general not correspond to no-trade equilibria. As an example consider the following specification of the three generations model with loglinear utility functions

$$U(c_0, c_1, c_2) = \ln c_0 + \ln c_1 + \ln c_2.$$

The demand functions of a generation born in period t are

$$\begin{aligned} d_0(\rho_t, \rho_{t+1}, \mathbf{w}) &= \frac{1}{3}\left[w_0 + \frac{1}{\rho_t} w_1 + \frac{1}{\rho_t \rho_{t+1}} w_2\right], \\ d_1(\rho_t, \rho_{t+1}, \mathbf{w}) &= \frac{1}{3}\left[\rho_t w_0 + w_1 + \frac{1}{\rho_{t+1}} w_2\right], \\ d_2(\rho_t, \rho_{t+1}, \mathbf{w}) &= \frac{1}{3}\left[\rho_t \rho_{t+1} w_0 + \rho_{t+1} w_1 + w_2\right], \end{aligned}$$

Condition (3.2) can then be written as[4]

$$\begin{aligned} &(2 - \rho_{t-1} - \rho_{t-1}\rho_{t-2})\, w_0 + \left(2 - \rho_{t-1} - \frac{1}{\rho_t}\right) w_1 \\ &\quad + \left(2 - \frac{1}{\rho_t} - \frac{1}{\rho_t \rho_{t+1}}\right) w_2 = 0. \end{aligned}$$

The steady state equilibria are the solutions of a fourth order polynomial. One of the roots is $\rho^* = 1$, which corresponds to the monetary steady state. The other steady state equilibria are solutions of

[3]The autarkic steady state interest factor ρ^a is then such that it is optimal for an agent to consume his own endowment. It is easily seen to be

$$\rho^a = \frac{p_t}{p_{t+1}} = \frac{U_1(w_0, w_1)}{U_2(w_0, w_1)},$$

where $U_i(.,.)$ is the partial derivative of the utility function with respect to the i'th variable.

[4]For this model the temporal equilibrium map F can be explicitly derived as

$$\rho_{t+1} = \frac{w_2}{\rho_t\left[(2 - \rho_{t-1} - \rho_{t-1}\rho_{t-2})\, w_0 + \left(2 - \rho_{t-1} - \frac{1}{\rho_t}\right) w_1 + \left(2 - \frac{1}{\rho_t}\right) w_2\right]}.$$

$$w_0\rho^3 + (2w_0 + w_1)\rho^2 - (w_1 + 2w_2)\rho - w_2 = 0. \tag{3.4}$$

It can easily be checked that (3.4) has at most one positive root. Notice that if an agent only has endowment when young, or only when old, a balanced steady state interest factor does not exist. Now we briefly consider two examples of balanced equilibria.

First let endowments be $\mathbf{w} = (8, 12, 1)$. The unique balanced steady state then corresponds to $\rho^b = \frac{1}{2}$. Individual excess demands are

$$\begin{aligned}
d_0\left(\rho^b, \rho^b, \mathbf{w}\right) - w_0 &= \frac{1}{3}\left[w_0 + 2w_1 + 4w_2\right] - w_0 = 12 - 8 = 4 \\
d_1\left(\rho^b, \rho^b, \mathbf{w}\right) - w_1 &= \frac{1}{3}\left[\frac{1}{2}w_0 + w_1 + 2w_2\right] - w_1 = 6 - 12 = -6 \\
d_2\left(\rho^b, \rho^b, \mathbf{w}\right) - w_2 &= \frac{1}{3}\left[\frac{1}{4}w_0 + \frac{1}{2}w_1 + w_2\right] - w_2 = 3 - 1 = 2
\end{aligned}$$

Trade proceeds as follows: the young generation buys 4 units from the middle aged and pays $\frac{1}{2} \times 4 = 2$ back when middle aged itself to the then old generation.

As another example consider the endowment vector $\mathbf{w} = (5, 0, 16)$. The steady state interest factor is $\rho^b = 2$. Individual excess demands are

$$\begin{aligned}
d_0\left(\rho^b, \rho^b, \mathbf{w}\right) - w_0 &= \frac{1}{3}\left[w_0 + \frac{1}{2}w_1 + \frac{1}{4}w_2\right] - w_0 = 3 - 5 = -2 \\
d_1\left(\rho^b, \rho^b, \mathbf{w}\right) - w_1 &= \frac{1}{3}\left[2w_0 + w_1 + \frac{1}{2}w_2\right] - w_1 = 6 - 0 = 6 \\
d_2\left(\rho^b, \rho^b, \mathbf{w}\right) - w_2 &= \frac{1}{3}\left[4w_0 + 2w_1 + w_2\right] - w_2 = 12 - 16 = -4
\end{aligned}$$

The young generation trades 2 units of the commodity with the middle-aged consumer for $2 \times 2 = 4$ units in the next period.

For both examples consumption in the monetary steady state is $(c_0^*, c_1^*, c_2^*) = (7, 7, 7)$. It can easily be checked that for the first example aggregate savings are positive in the monetary steady state (the Samuelson case) and in the second example aggregate savings are negative in the monetary steady state (the classical case). Gale (1973) shows that for the Samuelson (classical) case one must have $\rho^b < \rho^*$ ($\rho^b > \rho^*$). The intuition is clear: in order to give up saving in the Samuelson case, interest rates have to be negative and in order to give up borrowing in the classical case interest rates have to be positive.

Optimality of the balanced equilibria has received a lot of attention in the literature. Balanced equilibria in the Samuelson case are not Pareto optimal. From the first example above this can be seen by letting each middle-aged agent transfer some endowment to the then old. Utility then increases for each generation. For the second example it is not possible to arrive at a Pareto improvement by letting each old agent transfer some of his income to the middle aged, because this makes the old agent of period 0 worse off.

Up to now we have considered only stationary equilibrium paths. We could also have equilibrium paths where prices move in a cycle, for example when a price path of the form $(\ldots, p_0, p_1, p_2, p_0, p_1, p_2, p_0, \ldots)$ satisfies condition (3.1) for all t. For some time it has been known that the overlapping generations model can exhibit these perfect foresight equilibrium cycles (see e.g. Benhabib and Day (1982) and Grandmont (1985)). There is a correspondence between these equilibrium cycles and equilibria in cyclical general equilibrium models. Different generations in the overlapping generations model have the same preferences and endowments but differ in the prices they face. Different agents in an exchange economy face the same prices but differ in their preferences and endowments. In this chapter we investigate the existence of (monetary) cycles of increasing length in overlapping generations models with a given number of generations, by utilizing the correspondence between these cycles and the equilibria in a sequence of cyclical general equilibrium models, with an increasing number of agents and commodities. Such a correspondence was, as far as we know, first mentioned in Balasko and Ghiglino (1995) for the case of 2-cycles in two generations models. We study standard overlapping generations models with n generations, one commodity per period and no bequest motive, while all generations have the same preferences and endowments. We look at cycles of length $m \geq n$, with the help of general equilibrium models with m goods and m agents having cyclically permuted preferences in only n goods (and eventually a permuted production structure). Symmetric and asymmetric equilibria correspond to steady states and equilibrium cycles, respectively. It seems to be impossible to extend the equivalence result to balanced steady states or balanced cycles in overlapping generations models.

The rest of this chapter is organized as follows. In Section 3.2 we define the cyclical exchange economy and show that the asymmetric equilibria of this economy correspond to an equilibrium cycle in the overlapping generations model. Furthermore we show that this approach can be applied to two generations models to study the existence of cycles of any period and we show that the model can be extended in different ways. In Section 3.3 we give some examples. The first example is a three generations model where cycles of different periods are shown to exist. The second example is a two generations model where cycles of any period can be shown to exist and the last example is a two generations model with capital, where also cycles of any period can be shown to exist.

3.2 Equivalence

3.2.1 A cyclical exchange economy

In Chapter 2 we constructed exchange economies with three commodities and three consumers which had different kinds of symmetries: cyclical, reflectional or both. In this section we construct an exchange economy with $n+1$ agents and commodities that has a cyclical symmetry. This construction is equivalent to the one used in Proposition 2.1 of Chapter 2. We let $U{:}\mathbb{R}_+^{n+1} \to \mathbb{R}$ be a strictly quasi concave, strictly monotonic utility function and $\mathbf{p} = (p_0, ..., p_n) \in \mathbb{R}_+^{n+1}$ and $\mathbf{w} = (w_0, ..., w_n) \in \mathbb{R}_+^{n+1}$ a vector of prices and a vector of endowments respectively. The utility maximizing consumption bundle of an agent with these preferences and endowments is

$$\mathbf{x} = \mathbf{d}(\mathbf{p}, \mathbf{w}) = \arg \max_{\mathbf{x} \in \mathbb{R}_+^{n+1}} \{U(\mathbf{x}) \mid \mathbf{p} \cdot \mathbf{x} \leq \mathbf{p} \cdot \mathbf{w}\}, \tag{3.5}$$

where individual excess demand is given by $z(\mathbf{p}, \mathbf{w}) = \mathbf{d}(\mathbf{p}, \mathbf{w}) - \mathbf{w}$. We want to construct an exchange economy that has symmetry group M (that is, the aggregate excess demand functions must commute with M : $M \circ \mathbf{z} = \mathbf{z} \circ M$), where M is the permutation that shifts all elements of a vector one position backward and the first element to the last position:[5]

$$(y_0, y_1, ..., y_n) \xrightarrow{M} (y_1, ..., y_n, y_0). \tag{3.6}$$

The inverse M^{-1} of this permutation shifts all elements one position forward and the last element to the first position; applying M $n+1$ times leaves a vector invariant. M can be viewed as the following $(n+1) \times (n+1)$ matrix

[5]In the notation of Chapter 2 this permutation would be denoted $M_{23\cdots(n+1)1}$.

$$\begin{pmatrix} 0 & 1 & 0 & \cdots & 0 \\ 0 & 0 & 1 & \cdots & 0 \\ \vdots & \vdots & \vdots & \ddots & \vdots \\ 0 & 0 & 0 & \cdots & 1 \\ 1 & 0 & 0 & \cdots & 0 \end{pmatrix}$$

working on a column vector. Observe that $M' = M^{-1}$ and $M^{n+1} = I$.

Now we define a cyclical exchange economy with $n+1$ agents, by constructing agents from the utility function U and the vector of endowments $\mathbf{w}$ as follows.

Definition 3.1 *Given a utility function $U : I\!R_+^{n+1} \to I\!R$ and a vector of endowments $w = (w_0, ..., w_n) \in I\!R_+^{n+1}$ the corresponding cyclical exchange economy $\mathcal{E} = \{U, \mathbf{w}, n+1\}$ is the exchange economy consisting of $n+1$ agents where the i'th agent has utility function $U\left(M^i\mathbf{x}\right)$ and vector of endowments $M^{-i}w$, $i = 0, 1, ..., n$.*

The following proposition then shows that this cyclical exchange economy has symmetry M, as required.

Proposition 3.1 *The vector of aggregate excess demand functions $\mathbf{z}(\mathbf{p}) \equiv \sum_{i=0}^{n} z_i(\mathbf{p}, \mathbf{w})$ of $\mathcal{E}$ has* symmetry M, *that is*

$$\mathbf{z}(M\mathbf{p}) = M\mathbf{z}(\mathbf{p}).$$

Furthermore we have

- $\widehat{\mathbf{p}} = (1, 1, \ldots, 1)$ *is an equilibrium price vector*
- *If $\mathbf{p}^* \neq \widehat{\mathbf{p}}$ is an equilibrium price vector then so is $M^k\mathbf{p}^*$ for $k = 1, 2, \ldots, n$.*

Proof. First consider the problem of maximizing $U\left(M^i\mathbf{x}\right)$ subject to the budget constraint $\mathbf{p} \cdot \mathbf{x} \leq \mathbf{p} \cdot \left(M^{-i}\mathbf{w}\right)$. This is equivalent to maximizing $U\left(M^i\mathbf{x}\right)$ subject to $\left(M^i\mathbf{p}\right) \cdot M^i\mathbf{x} \leq \left(M^i\mathbf{p}\right) \cdot \mathbf{w}$. The solution of this last problem satisfies $M^i\mathbf{x} = \mathbf{d}\left(M^i\mathbf{p}, \mathbf{w}\right)$ so the utility maximizing consumption bundle for the i'th agent given prices and endowments is

$$\mathbf{x}^i(\mathbf{p}) = M^{-i}\mathbf{d}\left(M^i\mathbf{p}, \mathbf{w}\right) = \arg\max_{\mathbf{x} \in I\!R_+^{n+1}} \left\{U\left(M^i\mathbf{x}\right) \mid \mathbf{p} \cdot \mathbf{x} \leq \mathbf{p} \cdot \left(M^{-i}\mathbf{w}\right)\right\}, \tag{3.7}$$

where

$$x_j^i(\mathbf{p}) = d_{j-i}\left(M^i\mathbf{p}, \mathbf{w}\right) \qquad i, j = 0, 1, \ldots, n,$$

is the consumption of commodity j by agent i. Now $\mathbf{x}(\mathbf{p}) = \sum_{i=0}^{n} \mathbf{x}^i(\mathbf{p})$ can be written as

$$\mathbf{x}(\mathbf{p}) = \begin{pmatrix} d_0(\mathbf{p}, \mathbf{w}) \\ d_1(\mathbf{p}, \mathbf{w}) \\ \vdots \\ d_n(\mathbf{p}, \mathbf{w}) \end{pmatrix} + \begin{pmatrix} d_n(M\mathbf{p}, \mathbf{w}) \\ d_0(M\mathbf{p}, \mathbf{w}) \\ \vdots \\ d_{n-1}(M\mathbf{p}, \mathbf{w}) \end{pmatrix} + \ldots + \begin{pmatrix} d_1(M^n\mathbf{p}, \mathbf{w}) \\ d_2(M^n\mathbf{p}, \mathbf{w}) \\ \vdots \\ d_0(M^n\mathbf{p}, \mathbf{w}) \end{pmatrix}. \tag{3.8}$$

By (3.7)

$$M\mathbf{x}^i(\mathbf{p}) = M^{1-i}\mathbf{d}\left(M^i\mathbf{p}, \mathbf{w}\right) = M^{1-i}\mathbf{d}\left(M^{i-1}(M\mathbf{p}), \mathbf{w}\right) = \mathbf{x}^{i-1}(M\mathbf{p})$$

that is: the demand of agent $i-1$ at prices $M\mathbf{p}$ equals $M\mathbf{x}^i(\mathbf{p})$. Hence, given the $n+1$ tuple of demand vectors $\left(\mathbf{x}^0(\mathbf{p}), \mathbf{x}^1(\mathbf{p}), \ldots, \mathbf{x}^n(\mathbf{p})\right)$ at $\mathbf{p}$, $\left(M\mathbf{x}^0(\mathbf{p}), M\mathbf{x}^1(\mathbf{p}), .., M\mathbf{x}^n(\mathbf{p})\right) = \left(\mathbf{x}^n(M\mathbf{p}), \mathbf{x}^0(M\mathbf{p}), .., \mathbf{x}^{n-1}(M\mathbf{p})\right)$ and it follows

$$\begin{aligned} \mathbf{z}(M\mathbf{p}, \mathbf{w}) &= \sum_{i=0}^{n} \mathbf{z}^i(M\mathbf{p}, \mathbf{w}) = \sum_{i=0}^{n} \left(\mathbf{x}^i(M\mathbf{p}) - M^{-i}\mathbf{w}\right) \\ &= \sum_{i=0}^{n} \left(M\mathbf{x}^{i+1}(\mathbf{p}) - M^{-i}\mathbf{w}\right) \\ &= M\sum_{i=0}^{n} \mathbf{z}^i(\mathbf{p}, \mathbf{w}) = M\mathbf{z}(\mathbf{p}, \mathbf{w}) \end{aligned}$$

using $\sum_{i=1}^{n} M^{-i}\mathbf{w} = \left(\sum_{j=0}^{n} w_j, \sum_{j=0}^{n} w_j, \ldots, \sum_{j=0}^{n} w_j\right)' = \sum_{i=0}^{n} M^{1-i}\mathbf{w}$ and $\mathbf{x}^{n+1}(\mathbf{p}) = \mathbf{x}^0(\mathbf{p})$. Denote $\mathbf{e} = (1, 1, \ldots, 1)$. For all j, k $z_j(\mathbf{e}) = z_k(\mathbf{e})$ while

$$z_j(\mathbf{e}) = \sum_{i=0}^{n} d_i(\mathbf{e}, \mathbf{w}) - \sum_{i=0}^{n} w_i = 0$$

coincides with the budget equation for each agent. Furthermore, if $\mathbf{p}^*$ is an equilibrium then by symmetry, so is $M^k\mathbf{p}^*$ since

$$\mathbf{z}\left(M^k\mathbf{p}^*\right) = M^k\mathbf{z}(\mathbf{p}^*) = 0.$$

■

The equilibrium price vector $\mathbf{p}^*$ might consist of a repetition of a shorter price vector with length m such that $km = n+1$, where k is an integer. In this case we have $M^{m+1}\mathbf{p}^* = \mathbf{p}^*$ and the number of asymmetric equilibria corresponding to $\mathbf{p}^*$ then is m. Given that generically the number of equilibria in an exchange economy is odd (Dierker (1972)), this implies that for m even (odd) there must be an even (including zero) (odd) number of other asymmetric equilibria. In particular, if the number of commodities is odd and there are multiple equilibria, then there have to be at least two different asymmetric price vectors not related to each other by symmetry (compare the first example of Section 3.3).

3.2.2 An overlapping generations model

We now present an overlapping generations model with $n+1$ generations, similar to the model with three generations discussed in Section 3.1. Consider an overlapping generations model (denoted $\mathcal{OG}$) with $n+1$ generations and stationary endowments and preferences. Let each generation consist of a single agent. A generation born in period t has endowments $\mathbf{w} = (w_0, w_1, \ldots, w_n)$ and utility function $U(\mathbf{c}^t)$ with $\mathbf{c}^t = (c_0^t, \ldots, c_n^t)$, where c_s^t denotes consumption in period $t+s$ of a person born in period t. The model is specified by $\mathcal{OG} = \{U, \mathbf{w}, n+1\}$. A generation born in period t faces prices $\mathbf{p}^t = (p_t, \ldots, p_{t+n})$ and maximizes $U(\mathbf{c}^t)$ subject to his budget constraint $\mathbf{p}^t \cdot \mathbf{c}^t \leq \mathbf{p}^t \cdot \mathbf{w}$. This gives a vector of demand functions

$$\mathbf{c}^t = \mathbf{d}\left(\mathbf{p}^t, \mathbf{w}\right), \tag{3.9}$$

or

$$c_s^t = d_s\left(\mathbf{p}^t, \mathbf{w}\right), \qquad s = 0, 1, \ldots, n.$$

In every period total demands of all living generations have to equal total endowments of all living generations. This gives the following set of equilibrium conditions (compare (3.1))

$$\sum_{s=0}^{n} c_s^{t-s} = \sum_{s=0}^{n} d_s\left(\mathbf{p}^{t-s}, \mathbf{w}\right) = \sum_{s=0}^{n} w_s \qquad \forall t. \tag{3.10}$$

A price path $\mathbf{p}^{\infty} = (\ldots, p_{-1}, p_0, p_1, p_2, \ldots)$ is an equilibrium path under perfect foresight if it satisfies conditions (3.10).

Consider equilibrium cycles of period $n+1$, that is price paths that repeat after $n+1$ periods: $\mathbf{p}^{\infty} = (\ldots, p_0, p_1, \ldots, p_n, p_0, p_1, \ldots, p_n, p_0, \ldots)$. For $\mathbf{p} \equiv \mathbf{p}^0 = (p_0, p_1, \ldots, p_{n-1}, p_n)$ we have $\mathbf{p}^t = M^t \mathbf{p}$, hence (3.10) becomes

$$\sum_{s=0}^{n} c_s^{t-s} = \sum_{s=0}^{n} d_s \left(M^{t-s}\mathbf{p}, \mathbf{w}\right) = \sum_{s=0}^{n} w_s \qquad t = 0, 1, \ldots, n. \tag{3.11}$$

We denote by $\mathcal{OG}(n+1)$ the model $\mathcal{OG}$ with equilibrium conditions (3.11).

3.2.3 Equivalence of cycles and asymmetric equilibria

In this section we will show that there is a one-to-one correspondence between the equilibrium cycles of the overlapping generations model and (asymmetric) equilibria of the cyclical exchange economy defined above.

Theorem 3.1 *$\mathcal{OG}(n+1)$ has equilibrium cycle p^* if and only if the cyclical exchange economy $\mathcal{E}$ has equilibrium price vector p^*.*

Proof. Using $M = M^{-n}, M^2 = M^{1-n}$ etc. we can rewrite (3.8) as

$$\mathbf{x}(\mathbf{p}) = \begin{pmatrix} d_0(\mathbf{p}, \mathbf{w}) + d_1(M^{-1}\mathbf{p}, \mathbf{w}) + \ldots + d_n(M^{-n}\mathbf{p}, \mathbf{w}) \\ d_0(M^{-n}\mathbf{p}, \mathbf{w}) + d_1(\mathbf{p}, \mathbf{w}) + \ldots + d_n(M^{1-n}\mathbf{p}, \mathbf{w}) \\ \vdots \\ d_0(M^{-1}\mathbf{p}, \mathbf{w}) + d_1(M^{-2}\mathbf{p}, \mathbf{w}) + \ldots + d_n(\mathbf{p}, \mathbf{w}) \end{pmatrix}.$$

From this it follows that an equilibrium price vector $\mathbf{p}^*$ of $\mathcal{E}$ has to satisfy

$$\sum_{i=0}^{n} d_i \left(M^{j-i}\mathbf{p}^*, \mathbf{w}\right) = \sum_{i=0}^{n} w_i, \qquad j = 0, 1, ..., n. \tag{3.12}$$

By (3.11) the equilibrium conditions in $\widehat{\mathcal{OG}}(n+1)$ are

$$\sum_{s=0}^{n} d_s \left(M^{t-s}\mathbf{p}^*, \mathbf{w}\right) = \sum_{s=0}^{n} w_s, \qquad t = 0, 1, ..., n. \tag{3.13}$$

Conditions (3.12) and (3.13) are exactly the same. So the existence of an equilibrium cycle $\mathbf{p}^*$ in $\widehat{\mathcal{OG}}(n+1)$ is equivalent with the existence of an equilibrium price vector $\mathbf{p}^*$ in $\mathcal{E}$. ■

By Proposition 3.1, $\mathbf{p} = (1, 1, \ldots, 1)$ is an equilibrium price in $\mathcal{E}$, hence by Theorem 3.1 it is an equilibrium in $\mathcal{OG}(n+1)$, the existence of which is well known and easily proved directly from (3.10) or (3.11). The corresponding equilibrium allocation in $\mathcal{OG}(n+1)$ is usually called the monetary steady state.

Theorem 3.1 also says that if $\mathcal{E}$ has an asymmetric equilibrium $\mathbf{p}$, then $\mathcal{OG}(n+1)$ has equilibrium cycle $\mathbf{p}$. Its period is equal to m (with $n+1 = km$, where k is an integer), if $\mathbf{p}$ is a repetition of k identical m-cycles. Theorem 3.1 therefore only identifies cycles with a period which is equal to $\frac{n+1}{k}$ times the number of generations in $\mathcal{OG}$.

An extension to cycles with a period different from $n+1$, possibly consisting of subcycles, is however straightforward. Let

$$\widehat{U}(c_0, c_1, \ldots, c_n, c_{n+1}, \ldots, c_{n'}) \equiv U(c_0, \ldots, c_n),$$

and

$$\widehat{\mathbf{w}} \equiv (w_0, \ldots, w_n, 0, \ldots, 0),$$

then $\widehat{\mathcal{E}}(n'+1) = \left\{\widehat{U}, \widehat{\mathbf{w}}, n+1, n'+1\right\}$ is again a cyclical exchange economy. A consumer in $\widehat{\mathcal{E}}(n'+1)$ only consumes and owns the first $n+1$ commodities (in the order in which they appear in his own utility function).

Proposition 3.2 *$\mathcal{OG}(n+1)$ has equilibrium cycle p^* of length $n'+1$ if and only if $\widehat{\mathcal{E}}(n'+1)$ has an asymmetric equilibrium price vector.*

Proof. Demands for agent 0 are

$$\tilde{d}_j(q_0, q_1 \ldots, q_{n'}, \widehat{\mathbf{w}}) = \begin{cases} d_j(q_0, \ldots, q_n, \mathbf{w}) & j = 0, 1, \ldots, n \\ 0 & j = n+1, \ldots, n' \end{cases},$$

where $\mathbf{q} \in I\!R_+^{n'+1}$ is a price vector. Agent i in $\widehat{\mathcal{E}}(n'+1)$ then has demand function

$$\widehat{d}_j(\mathbf{p}, \widehat{\mathbf{w}}) = \tilde{d}_{j-i}\left(\widehat{M}^i\mathbf{p}, \widehat{\mathbf{w}}\right),$$

where $\widehat{M}$ is a permutation matrix of order $n'+1$. In $\mathcal{OG}$ it can be assumed without loss of generality, that every agent lives for $n'+1$ periods, but consumes and owns resources only in the first $n+1$ periods of his life. The demand functions of an agent born in period t are

$$\widehat{d}_s(p_t,\dots,p_{t+n'},\widehat{\mathbf{w}}) = \begin{cases} d_s(p_t,\dots,p_{t+n},\mathbf{w}) & s=0,1,\dots,n \\ 0 & s=n+1,\dots,n' \end{cases}$$

In every period there are $n'+1$ different generations, but only $n+1$ of these contribute to the economy in the sense that they may have a nonzero excess demand for consumption. Equilibrium cycles with period $n'+1$ must satisfy (3.10) but with $t=0,1,\dots,n'$, since they repeat after $n'+1$ periods. Denote by $\widehat{\mathcal{OG}}(n'+1) = \left\{\widehat{U},\widehat{\mathbf{w}},n+1,n'+1\right\}$ the overlapping generations model with equilibrium cycles specified by

$$\sum_{s=0}^{n'} \widehat{d}_s\left(\widehat{M}^{t-s}\mathbf{p},\widehat{\mathbf{w}}\right) = \sum_{s=0}^{n'} \widehat{w}_s, \qquad t=0,1,\dots,n'.$$

Note that $\mathcal{OG}(n+1)$ and $\widehat{\mathcal{OG}}(n'+1)$ are special cases of $\mathcal{OG}$ and that these three models are not qualitatively different, since equilibrium cycles in all three satisfy (3.10). By Theorem 3.1, $\mathbf{p}$ is an equilibrium cycle in $\widehat{\mathcal{OG}}(n'+1)$ if and only if $\mathbf{p}$ is an equilibrium in $\widehat{\mathcal{E}}(n'+1)$ .Now assume $\widehat{\mathcal{E}}(n'+1)$ has an asymmetric equilibrium. We consider two cases. If this asymmetric equilibrium price vector does not consist of a repetition of a shorter price vector, by Theorem 3.1, $\widehat{\mathcal{OG}}(n'+1)$ has a $n'+1$ cycle. Therefore also $\mathcal{OG}$ has a $n'+1$ cycle. If, on the other hand, the asymmetric equilibrium price vector consists of a repetition of subvectors of length m, where $m < n+1$ we find that $\widehat{\mathcal{OG}}(n'+1)$ and therefore also $\mathcal{OG}$ have a m-cycle. ∎

By looking at a series of exchange economies $\widehat{\mathcal{E}}(n'+1)$ we can try to find equilibrium cycles of any period of the overlapping generations model. Notice that it is not the number of generations of the overlapping generations model, but the length of the equilibrium cycle that determines the number of agents and commodities in the exchange economy we need to study. If, for example, an overlapping generations model has two coexisting equilibrium cycles of different length, then these equilibrium cycles correspond to the asymmetric equilibria of two *different* cyclical exchange economies, $\widehat{\mathcal{E}}(n'+1)$ and $\widehat{\mathcal{E}}(n''+1)$.

3.2.4 A special case: the two generations overlapping generations model and Sarkovskii's theorem

We want to spend some extra space on the overlapping generations model with two generations. The equilibrium condition for this overlapping generations model is

$$(d_0(p_t, p_{t+1}, \mathbf{w}) - w_0) + (d_1(p_{t-1}, p_t, \mathbf{w}) - w_1) = 0,$$

which can be written as a dynamical system implicitly as

$$Z_0(\rho_t) = -Z_1(\rho_{t-1}) \tag{3.14}$$

where $\rho_t \equiv \frac{p_t}{p_{t+1}}$ is the interest factor in period t. As was discussed in Section 3.1, this dynamical system has generically two steady states, the monetary steady state $\rho^* = 1$ and the autarkic steady state ρ^a for which we have $Z_0(\rho^a) = Z_1(\rho^a) = 0$ and which does not correspond to an equilibrium in an exchange economy. We can distinguish two cases: the *classical case* where $\rho^a > \rho^*$ and where in the monetary equilibrium people borrow when they are young and pay back when they are old and the *Samuelson case* where $\rho^a < \rho^*$ and where in the monetary equilibrium people save when they are young and dissave when they are old. It is well-known that the autarkic steady state is unstable in the classical case and locally stable in the Samuelson case.[6] For the overlapping generations model with two generations we can use our methods in a special way. This is due to the following well-known result from the theory on nonlinear dynamical systems (Devaney (1989), p.62)

Theorem 3.2 ***(Sarkovskii's Theorem)*** *Consider the following ordering of all positive integers:*

$$\begin{aligned} 3 &\succ 5 \succ 7 \succ \ldots \succ 2\cdot 3 \succ 2\cdot 5 \succ 2\cdot 7 \succ \ldots \\ &\succ 2^k\cdot 3 \succ 2^k\cdot 5 \succ 2^k\cdot 7 \succ \ldots \succ 2^2 \succ 2 \succ 1 \end{aligned}$$

---

[6] That this is so can be easily seen by differentiating the budget restriction of the consumer at ρ^a to obtain

$$-\frac{Z_1'(\rho^a)}{Z_0'(\rho^a)} = \rho^a.$$

Then linearizing (3.14) around the autarkic steady state gives

$$(\rho_t - \rho) = -\frac{Z_1'(\rho^a)}{Z_0'(\rho^a)}(\rho_{t-1} - \rho) = \rho^a(\rho_{t-1} - \rho).$$

If f is a continuous map of an interval into itself with a periodic point of period p and $p \succ q$ in this ordering, then f has a periodic point of period q.

From this theorem it follows that if the overlapping generations model with two generations can be described as a continuous map that maps an interval into itself and that if this overlapping generations model has a period three cycle, it has cycles of every period.

Proposition 3.3 *The overlapping generations model with two generations can be represented by a uniquely defined, continuous function $f : I \to I$, if*

- $\rho^a > 1$ *and* $Z_0'(\rho) \neq 0$*, for all* $\rho \in I = [0, \rho^a]$ *or*
- $\rho^a < 1$ *and* $Z_1'(\rho) \neq 0$ *for all* $\rho \in I = [0, \overline{\rho}]$*, where* $\overline{\rho} \in (\rho^a, 1]$ *is the largest interest factor for which*
$$Z_0(\rho) = -Z_1(\overline{\rho})$$
has a solution in $(\rho^a, 1)$.

Furthermore, if the overlapping generations model has a monetary cycle of period three, it has monetary cycles of any period.

Proof. $Z_1(\rho)$ and $Z_0(\rho)$ are continuously differentiable by the assumptions on the utility functions. First consider the classical case where $\rho^a > 1$. Since ρ^a is the only solution to $Z_0(\rho^a) + Z_1(\rho) = 0$ we know that if a function of the overlapping generations model exists, it maps $[0, \rho^a]$ into $[0, \rho^a]$. Furthermore, if $Z_0'(\rho) \neq 0$ on $[0, \rho^a]$ then we can invert (3.14) and write

$$\rho_t = Z_0^{-1}(-Z_1(\rho_{t-1})) = f(\rho_{t-1}).$$

Now consider the Samuelson case where $\rho^a < 1$. In this case we have to study the backward dynamics. Clearly this backward dynamic map g maps $[0, \overline{\rho}]$ into $[0, \overline{\rho}]$. Furthermore, since the global inverse of $Z_1(\rho)$ exists, we have

$$\rho_{t-1} = Z_1^{-1}(-Z_0(\rho_t)) = g(\rho_t)$$

which is continuous. This completes the first statement. Application of Sarkovskii's theorem then gives the last part. ■

This proposition also implies that if the overlapping generations model with two generations has a three cycle we can construct a series of exchange economies $\widehat{\mathcal{E}}(n' + 1)$, having multiple equilibria for any $n' \geq 1$.

3.2.5 Extensions

We have considered, except for the number of generations, the most basic overlapping generations model with only one commodity, one agent per generation and a constant population. The model can be extended to allow for more commodities and more agents per generations and for a growing population. This section briefly deals with these extensions.

More goods or more consumers

The case with m goods, but with a single consumer in each period, is straightforward. The previous analysis can be repeated, but with $\mathbf{x} = (x_0, x_1, \ldots, x_n)$ replaced by $\mathbf{x} = (\mathbf{x}_0, \mathbf{x}_1, \ldots, \mathbf{x}_n)$, with $\mathbf{x}_s = (x_{s1}, x_{s2}, \ldots, x_{sm})$; similarly for prices $\mathbf{p}$ and resources $\mathbf{w}$. The transformation M now works on vectors: $M(\mathbf{x}_0, \mathbf{x}_1, \ldots, \mathbf{x}_n) = (\mathbf{x}_1, \mathbf{x}_2, \ldots, \mathbf{x}_n, \mathbf{x}_0)$. From the utility function $U(\mathbf{x}_0^t, \mathbf{x}_1^t, \ldots, \mathbf{x}_n^t)$ and the resource vector $\mathbf{w}$, $(n+1)\times m$ demand functions are derived (compare (3.5))

$$x_{sj} = d_{sj}(\mathbf{p}, \mathbf{w}), \qquad s = 0, 1, \ldots, n, \quad j = 1, 2, \ldots, m. \tag{3.15}$$

$\mathcal{OG}$ now is an overlapping generations economy with m goods in each period. The equilibrium conditions (3.11) for a cycle of period $n+1$ are replaced by

$$\sum_{s=0}^{n} \mathbf{c}_s^{t-s} = \sum_{s=0}^{n} \mathbf{d}_s(M^{t-s}\mathbf{p}, \mathbf{w}) = \sum_{s=0}^{n} \mathbf{w}_s, \qquad t = 0, 1, \ldots, n. \tag{3.16}$$

The agents of the corresponding symmetric exchange economy $\mathcal{E}(n+1)$ are specified by $U(M^i\mathbf{x})$ and $M^{-i}\mathbf{w}$ $(i = 0, 1, \ldots n)$ (compare Definition 3.1). Theorem 3.1 remains valid: in the proof scalar functions d_s are replaced by vector functions $\mathbf{d}_s$.

Note that in this case there exists $\overline{\mathbf{p}} = (\overline{p}_1, \ldots, \overline{p}_m)$, such that $\mathbf{p}^0 = (\overline{\mathbf{p}}, \overline{\mathbf{p}}, \ldots, \overline{\mathbf{p}})$ is an equilibrium of the exchange economy. Maximize $U(\mathbf{x})$ under the constraint $\sum_{t=0}^{n} \mathbf{x}^t \leq \sum_{t=0}^{n} \mathbf{w}^t \equiv \mathbf{W}$. Since the constraint set is convex and compact, a solution $\mathbf{x}^*$ exists. Substitution of the constraint for $\mathbf{x}^0$ in U gives

$$V(\mathbf{x}^1, \mathbf{x}^2, \ldots, \mathbf{x}^n) = U(\mathbf{W} - \mathbf{x}^1 - \mathbf{x}^2 \cdots - \mathbf{x}^n, \mathbf{x}^1, \ldots, \mathbf{x}^n).$$

Differentiating in $\mathbf{x}^*$ gives the first order conditions

$$\frac{\partial V}{\partial x_i^k} = -\frac{\partial U}{\partial x_i^0} + \frac{\partial U}{\partial x_i^k} = 0, \qquad i = 1, 2, \ldots, m, \ k = 1, 2, \ldots, n.$$

If we set $\overline{p}_i = \partial U/\partial x_i^0$, then $\mathbf{x}^* = \mathbf{d}(\overline{\mathbf{p}}, \overline{\mathbf{p}}, \ldots, \overline{\mathbf{p}})$ maximizes U under $\overline{\mathbf{p}} \cdot \left(\sum_{t=0}^{n} \mathbf{x}^t - \sum_{t=0}^{n} \mathbf{w}^t\right) \leq 0$, which implies $\mathbf{x}^*$ is an equilibrium at $\mathbf{p}^0 = (\overline{\mathbf{p}}, \overline{\mathbf{p}}, \ldots, \overline{\mathbf{p}})$: it represents the "monetary" steady state in this economy.

With l consumers $h = 1, 2, \ldots, l$, in each period and a single good, we must have l utility functions $U^h\left(\mathbf{x}^h\right)$ and l resource vectors $\mathbf{w}^h = (w_0^h, w_1^h, \ldots, w_n^h)$ giving $n+1$ demand functions for each agent h, giving, instead of (3.15)

$$x_s^h = d_s^h\left(\mathbf{p}, \mathbf{w}^h\right), \qquad s = 0, 1, \ldots, n, \; h = 1, 2, \ldots, l.$$

$\mathcal{OG} = \left\{U^h, \mathbf{w}^h, n+1\right\}$ is now an economy with l consumers in each period. An equilibrium cycle of period $n+1$ is a solution of (instead of (3.10))

$$\sum_{h=1}^{l}\sum_{s=0}^{n} c_s^{h,t-s} = \sum_{h=1}^{l}\sum_{s=0}^{n} d_s^h\left(M^{t-s}\mathbf{p}, \mathbf{w}^h\right) = \sum_{h=1}^{l}\sum_{s=0}^{n} w_s^h, \qquad t = 0, 1, \ldots n. \tag{3.17}$$

The corresponding cyclical exchange economy $\mathcal{E}\left(U^h, \mathbf{w}^h, n+1\right)$, has $l \times (n+1)$ consumers with utilities $U^h\left(\mathbf{M}^i\mathbf{x}\right)$ and resources $M^{-i}\mathbf{w}^h$. It is equivalent to the $\mathcal{OG}$–model by Theorem 1: in the proof the individual demands d_s have to be replaced by aggregate demands $\sum_{h=1}^{l} d_s^h$ and w_s is replaced by $\sum_{h=1}^{l} w_s^h$. Again $\mathbf{p} = \mathbf{e}$ is an equilibrium price vector: (3.17) coincides with the *sum* of all budget equations.

For a model with m goods *and* l consumers we get similar results.

Growth

We study the model $\mathcal{OG}(n+1, \beta)$ which is the model $\mathcal{OG}(n+1)$ of Section 3.2.2, with a single good in each period, a lifetime resource vector $\mathbf{w}$ and with the utility function $U(\mathbf{x})$ for all agents of all generations, but with a growing population: normalizing the number of agents born in $t = 0$ at 1, β^t $(t \in (-\infty, \infty))$ are born in t, β being the rate of population growth.

It is well known that with population growth at a rate β the monetary steady state of $\mathcal{OG}(n+1)$ is replaced by a growth path with identical consumption of all generations, but with prices decreasing at the rate β^{-1}, where Samuelson (1958) called the rate $\beta - 1$ the biological interest rate.

The equilibrium condition (3.10) for period t is replaced by

$$\sum_{s=0}^{n} \beta^{-s} c_s^{t-s} = \sum_{s=0}^{n} \beta^{-s} d_s \left(\mathbf{p}_{t-s}, \mathbf{w}\right) = \sum_{s=0}^{n} \beta^{-s} w_s. \tag{3.18}$$

$\mathbf{p} = \left(1, \beta^{-1}, \beta^{-2}, \beta^{-3}, ...\right)$ is always a solution, since for every t condition (3.18) then reduces to the budget equation of the individual agent. This price sequence corresponds to the monetary steady state $\rho = \frac{p_t}{p_{t+1}} = \beta$. Cycles are also possible. A cycle then has the following structure $\left(\ldots, \beta^{n+1} p_0, \beta^{n+1} p_1, \ldots, \beta^{n+1} p_n, p_0, \ldots, p_n, \beta^{-n-1} p_0, \ldots\right)$. Prices decrease along the cycle but interest factors $\rho_t = \frac{p_t}{p_{t+1}}$ move within a cycle. In fact, we have $\prod_{i=0}^{n} \rho_i = \beta^{n+1}$.

Now we construct $n+1$ utility functions and vectors of endowments in the following way: $V_i(\mathbf{x}) = U\left(M_\beta^i \mathbf{x}\right)$ and $\mathbf{w}_i = M_\beta^{-i} \mathbf{w}$, and $N_\beta = \left(M_\beta^{-1}\right)'$ where M_β and N_β are the following $(n+1) \times (n+1)$ matrices

$$M_\beta = \begin{pmatrix} 0 & 0 & \cdots & 0 & \beta^{-n} \\ \beta & 0 & \cdots & 0 & 0 \\ 0 & \beta & \cdots & 0 & 0 \\ \vdots & \vdots & \ddots & \vdots & \vdots \\ 0 & 0 & \cdots & \beta & 0 \end{pmatrix}, N_\beta = \begin{pmatrix} 0 & 0 & \cdots & 0 & \beta^{n} \\ \beta^{-1} & 0 & \cdots & 0 & 0 \\ 0 & \beta^{-1} & \cdots & 0 & 0 \\ \vdots & \vdots & \ddots & \vdots & \vdots \\ 0 & 0 & \cdots & \beta^{-1} & 0 \end{pmatrix}. \tag{3.19}$$

Notice that $M_\beta^{n+1} = I$ and $N_\beta = \left(M_\beta^{-1}\right)' = M_{\frac{1}{\beta}}$. The demand functions are

$$\mathbf{x}_i = M_\beta^{-i} \mathbf{d}\left(N_\beta \mathbf{p}, \mathbf{w}\right),$$

and the aggregate excess demand functions become

$$\mathbf{z}(\mathbf{p}) = \sum_{i=0}^{n} M_\beta^{-i} \mathbf{d}\left(N_\beta^i \mathbf{p}, \mathbf{w}\right) - \sum_{i=0}^{n} M_\beta^{-i} \mathbf{w}.$$

We have $\mathbf{z}(\mathbf{p}) =$

$$\begin{pmatrix} d_0(\mathbf{p},\mathbf{w}) + \beta^{-1}d_1(N_\beta\mathbf{p},\mathbf{w}) + . + \beta^{-n}d_n(N_\beta^n\mathbf{p},\mathbf{w}) - \sum_{i=0}^{n}\beta^{-i}w_i \\ \beta d_0(N_\beta^n\mathbf{p},\mathbf{w}) + d_1(\mathbf{p},\mathbf{w}) + . + \beta^{1-n}d_n\left(N_\beta^{n-1}\mathbf{p},\mathbf{w}\right) - \sum_{i=0}^{n}\beta^{1-i}w_i \\ \vdots \\ \beta^n d_0(N_\beta\mathbf{p},\mathbf{w}) + \beta^{n-1}d_1(N_\beta^2\mathbf{p},\mathbf{w}) + . + d_n(\mathbf{p},\mathbf{w}) - \sum_{i=0}^{n}\beta^{n-i}w_i \end{pmatrix}. \tag{3.20}$$

Multiplying the j'th element of $\mathbf{z}(\mathbf{p})$ by β^{-j} (where $j = 0, 1, \ldots, n$) gives the following $n+1$ equilibrium conditions

$$\sum_{i=0}^{n}\beta^{-i}d_i\left(N_\beta^{i-j}\mathbf{p},\mathbf{w}\right) = \sum_{i=0}^{n}\beta^{-i}w_i, \qquad j = 0, .., n+1. \tag{3.21}$$

Conditions (3.18) can be written in exactly the same way. The price vector $\widehat{\mathbf{p}} = \left(1, \beta^{-1}, \beta^{-2}, ..., \beta^{-n}\right)$ is always an equilibrium vector. Furthermore, from conditions (3.20) it follows that if $\mathbf{p}^*$ is an equilibrium, then so is $N_\beta^k\mathbf{p}^*$, for $k = 1, ..., n$. (Notice that $N_\beta\widehat{\mathbf{p}} = \widehat{\mathbf{p}}$.) Hence, if $\mathbf{p}^* = (p_0, p_1, \cdots, p_n)$ is an asymmetric equilibrium, then this exchange economy has n other asymmetric equilibria, i.e. $\left(\beta^{n+1}p_n, p_0, \cdots, p_{n-1}\right)$, $\left(\beta^{n+1}p_{n-1}, \beta^{n+1}p_n, p_0, \cdots, p_{n-2}\right), \cdots, \left(\beta^{n+1}p_1, \cdots, \beta^{n+1}p_n, p_0\right)$.

3.3 Examples

We will now illustrate the results obtained from the previous section by a set of examples.

3.3.1 A three generations overlapping generations model with CES utility functions

As a first example we study the occurrence of cycles in the overlapping generations model with three generations. A consumer born in period t has the CES utility function

$$U\left(c_t, c_{t+1}, c_{t+2}\right) = \left(c_t^{\theta} + \alpha_1 c_{t+1}^{\theta} + \alpha_2 c_{t+2}^{\theta}\right)^{\frac{1}{\theta}}, \qquad -\infty \leq \theta \leq 1, \alpha_1, \alpha_2 \geq 0, \tag{3.22}$$

and a vector of endowments $\mathbf{w} = (1, 0, 0)$. The individual demand functions are

$$\begin{aligned} d_0\left(p_t, p_{t+1}, p_{t+2}\right) &= \frac{p_t^{1-\sigma}}{p_t^{1-\sigma} + \alpha_1^{\sigma} p_{t+1}^{1-\sigma} + \alpha_2^{\sigma} p_{t+2}^{1-\sigma}}, \\ d_1\left(p_t, p_{t+1}, p_{t+2}\right) &= \frac{\alpha_1^{\sigma} p_t p_{t+1}^{-\sigma}}{p_t^{1-\sigma} + \alpha_1^{\sigma} p_{t+1}^{1-\sigma} + \alpha_2^{\sigma} p_{t+2}^{1-\sigma}}, \\ d_2\left(p_t, p_{t+1}, p_{t+2}\right) &= \frac{\alpha_2^{\sigma} p_t p_{t+2}^{-\sigma}}{p_t^{1-\sigma} + \alpha_1^{\sigma} p_{t+1}^{1-\sigma} + \alpha_2^{\sigma} p_{t+2}^{1-\sigma}}, \end{aligned} \tag{3.23}$$

where $\sigma = \frac{1}{1-\theta}$ is the elasticity of substitution. Recall that the corresponding cyclical exchange economy was also studied in Section 2.8 of Chapter 2. There we took $\sigma = \frac{1}{10}$ and $\alpha_1 = \alpha_2 = \alpha$ and studied the emergence of asymmetric equilibria as α decreased. It was found there that asymmetric equilibria appeared at about $\alpha \approx 0.051138$. Notice however that this exchange economy has two symmetries: a cyclical and a reflectional one. Here we want to study the occurrence of asymmetric equilibria when the exchange economy has only a cyclical symmetry. Therefore we take $\alpha_1 = \frac{1}{20}$ and $\alpha_2 = \frac{1}{25}$ and study the emergence of asymmetric equilibria as the elasticity of substitution decreases. This emergence of asymmetric equilibria occurs at $\sigma^* \approx 0.131$ (or $\theta^* \approx -6.634$). For $\sigma > \sigma^*$ the symmetric equilibrium is unique but for $\sigma < \sigma^*$ there are 7 different equilibrium price vectors. One of them is the symmetric equilibrium price vector, three of them correspond to one asymmetric equilibrium price vector and its permutations and three of them correspond to another asymmetric equilibrium price vector and its permutations. For example for $\sigma = \frac{1}{10}$ (or $\theta = -9$) the equilibria are $\widehat{\mathbf{p}} = (1, 1, 1)$, $\mathbf{p}^1 \approx (2.2637, 0.4121, 0.3242)$ and its permutations and $\mathbf{p}^2 \approx (2.2637, 0.4121, 0.3242)$ and its permutations. This means that for $\sigma = \frac{1}{10}$ $\mathcal{OG}$ has two 3-cycles in interest factors: $\boldsymbol{\rho}^1 \approx (5.493, 1.271, 0.143)$ and $\boldsymbol{\rho}^2 \approx (1.565, 1.261, 0.507)$.

We also study the occurrence of equilibrium cycles of period 2 and period 4 in the overlapping generations model $\widehat{\mathcal{OG}}\,(4)$. Let

$$V(x_0, x_1, x_2, x_3) = \left(x_0^\theta + \alpha_1 x_1^\theta + \alpha_2 x_2^\theta\right)^{\frac{1}{\theta}}, \qquad \widehat{\mathbf{w}} = (1, 0, 0, 0), \tag{3.24}$$

and construct the other three agents by permutation. Again consider $\alpha_1 = \frac{1}{20}$ and $\alpha_2 = \frac{1}{25}$. The symmetric equilibrium is unique for $\sigma \geq \sigma^0 \approx 0.374$. At σ^0 two new equilibria emerge of the form $\mathbf{p}^* = (p, q, p, q)$. Notice that such equilibria are not symmetric with respect to M, but they are symmetric with respect to M^2 (which is also a symmetry of $\widehat{\mathcal{E}}(4)$), since $M^2\mathbf{p}^* = \mathbf{p}^*$, $\mathbf{p}^*$ corresponds to a cycle of period 2 in $\mathcal{OG}$. As the elasticity of substitution decreases, 4 new equilibria of the form (p, q, r, s) emerge. These correspond to an equilibrium cycle of period 4 in $\mathcal{OG}$. This happens at $\sigma^1 \approx 0.190$.

3.3.2 A two generations overlapping generations model with CARA utility functions

Let an agent born in period t maximize the following *CARA* utility function[7]

$$U(x_0, x_1) = -e^{-x_0} - e^{-\beta x_1} \qquad 0 < \beta < 1, \tag{3.25}$$

where β is the *coefficient of absolute risk aversion* of the old generation. Furthermore we take $w_0 = w_1 = 1$. The demand functions of the corresponding (unpermuted) agent in the exchange economy are

$$d_0(p_0, p_1, \mathbf{w}) = \frac{\beta(p_0 + p_1) - p_1 \ln \beta \frac{p_0}{p_1}}{\beta p_0 + p_1} \text{ and} \tag{3.26}$$
$$d_1(p_0, p_1, \mathbf{w}) = \frac{p_0 + p_1 + p_0 \ln \beta \frac{p_0}{p_1}}{\beta p_0 + p_1}.$$

To study the existence of a 2-cycle we consider the corresponding economy $\mathcal{E}$. The excess demand function of the first commodity with

[7] This is the sum of two so called Constant Absolute Risk Aversion utility functions: for $U(c) = -e^{-\alpha c}$ the coefficient of absolute risk aversion $-\frac{U''(c)}{U'(c)}$ is equal to α and hence constant. Notice that this utility function is not homothetic.

commodity 2 as a numeraire equals

$$\begin{aligned} z_0(p,1) &= d_0(p,1,\mathbf{w}) + d_1(1,p,\mathbf{w}) - w_0 - w_1 \\ &= \frac{\beta(p+1) - \ln \beta p}{\beta p + 1} + \frac{1 + p + \ln \beta \frac{1}{p}}{\beta + p} - 2. \end{aligned}$$

Notice that $\lim_{p \to 0} z_0(p,1) > 0$ and $\lim_{p \to \infty} z_0(p,1) < 0$ and that $z_0(p,1)$ is continuous. This implies that if the excess demand function is upward sloping at the symmetric equilibrium $p^* = 1$, there have to be at least two other equilibria. The slope of the excess demand function is

$$\left.\frac{\partial z_0(p,1)}{\partial p}\right|_{p=1} = \frac{(\beta-1)\ln\beta - (\beta^2+3)}{(\beta+1)^2}.$$

So at $\beta^* \approx 0.04337$ two new equilibria emerge.[8] The two asymmetric equilibria correspond to a 2-cycle in $\mathcal{OG}$. For example: for $\beta = \frac{1}{25}$ we have the asymmetric equilibrium price vector $\mathbf{p}^* = (1.3442, 0.6558)$, so $\mathcal{OG}$ has a 2-cycle in interest factors: $\boldsymbol{\rho}^* = (2.0497, 0.4879)$. Hence the overlapping generations model undergoes a period doubling bifurcation at $\beta = \beta^*$. We can also look for 4-cycles in $\mathcal{OG}$ by studying $\widehat{\mathcal{E}}(4)$. For $\beta \in (0, \beta^0)$ there are, besides the symmetric equilibrium and the two other asymmetric equilibria, four asymmetric equilibria of the form $\mathbf{p} = (p,q,r,s)$. These equilibrium price vectors correspond to a 4-cycle in the overlapping generations model. So at β^0 a second period doubling bifurcation occurs. Using the LOCBIF program (Khibnik, Kuznetsov, Levitin and Nikolaev (1992)) we constructed Figure 3.1 for $\widehat{\mathcal{E}}(4)$.

Figure 3.1 shows what happens as β decreases: the cyclical exchange economy goes from a unique equilibrium to three equilibria to seven equilibria. The overlapping generations model with two generations therefore undergoes two period doubling bifurcations.

Equilibrium cycles with period three also occur. Let

$$W(x_0, x_1, x_2) = -e^{-x_0} - e^{-\beta x_1}, \quad \mathbf{w}^1 = (1,1,0)$$

[8] We would also have found these two extra equilibria if we would have replaced the CARA utility functions (3.25) by CES utility functions. This is similar to the result by Bala (1997) who finds for a cyclical exchange economy with two agents and CES utility functions that two extra asymmetric equilibria might emerge. These equilibria correspond to a period two cycle in the corresponding overlapping generations model. The approach by Bala differs from ours in that he does not seem to be aware of this correspondence and the fact that he is mainly interested in the dynamics of the continuous tâtonnement adjustment process, $\frac{\partial p_i}{\partial t} = z_0(p,1)$.

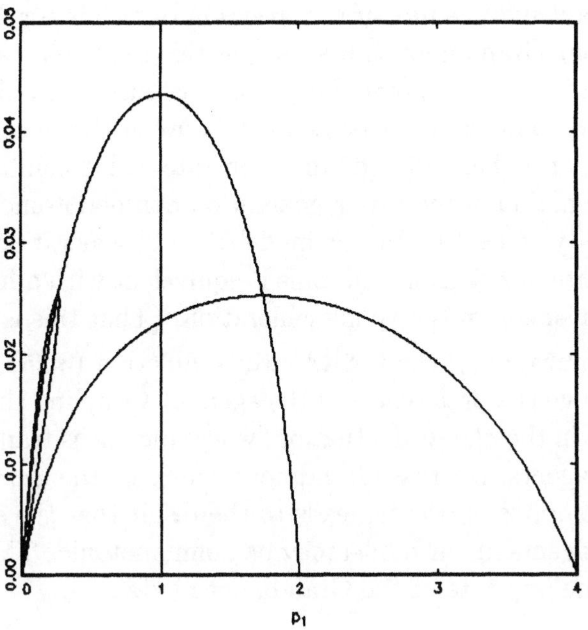

Figure 3.1: Equilibria in the cyclical exchange economy with four commodities and CARA utility functions as a function of β.

and construct the other 2 agents by permutation. $\widehat{\mathcal{E}}(3)$ has asymmetric equilibria for $\beta < \beta^c \approx 0.01013$. Hence $\mathcal{OG}$ with two generations and CARA utility functions has a 3-cycle for $\beta < \beta^c$. Furthermore, it is easily established that the conditions of Proposition 3.3 are satisfied so we immediately find

Proposition 3.4 *The overlapping generations model with two generations and CARA utility functions has equilibrium cycles of any period, for $\beta < \beta^c$. Furthermore, if we construct a cyclical exchange economy with $n' + 1$ commodities and $n' + 1$ agents by expanding and permuting the CARA utility functions and vector of endowments, with $\beta < \beta^c$, then this exchange economy has multiple equilibria for all $n' \geq 1$.*

The results obtained in our example correspond to the example in Grandmont (1985). The utility function he uses in his example is (see Grandmont (1985), p. 1029)

$$U(x_0, x_1) = 2\sqrt{x_0} + \frac{1}{1-\alpha} x_1^{1-\alpha}.$$

Here α is the constant *relative rate of risk aversion* (which is $-x_1U''/U'$) of the old generation. Grandmont shows that, as this relative rate of risk aversion increases to a certain point there is coexistence of equilibrium cycles of any period. This is due to a conflict between the income and substitution effects of a change in the inflation rate. This conflict may make the savings function of the young generation nonmonotonic, introducing the possibility of cycles. In our model the old generation has a very low *absolute* rate of risk aversion. This is equivalent with a high absolute rate of risk aversion of the young generation. (That this is so can easily be seen by multiplying the CARA utility function (3.25) by $e^{\frac{1}{\beta}}$, which does not change the preferences of the agents.) Coupling this with the fact that we are in the *classical* situation, where income is transferred from old to young agents, whereas Grandmont considers the *Samuelson* situation where the opposite occurs, leads to the result that the savings function of the old agents in our model may be nonmonotonic. Therefore our model is just the opposite of the Grandmont model.

3.3.3 An example with two generations and capital

Our results can also be applied to overlapping generations models with capital of the type studied by e.g. Reichlin (1986). An extension of Theorem 3.1 to these models with capital is straightforward. In this section we will give an example of such an overlapping generations model with capital that exhibits cycles of any period.

The commodity produced each period can be used for consumption and investment. The production technology is

$$y_t = f(k_{t-1}, l_t) + (1-\delta)\, k_{t-1} \tag{3.27}$$

where $f(.,.)$ gives net production and $(1-\delta)\, k_{t-1}$ gives the remaining capital stock after depreciation. Of course this setup requires that we have $k_t \geq (1-\delta)\, k_{t-1}$ for all t. We assume that f exhibits constant returns to scale with respect to capital and labour. Furthermore we assume labour is inelastically supplied and equal to 1. Profit for the firm in period t is then given by

$$p_t y_t - p_{t-1} k_{t-1} - q_t l_t$$

where q_t is the nominal wage rate. Capital demand follows from the first order condition

$$\frac{\partial f(k_{t-1}, 1)}{\partial k_{t-1}} + (1-\delta) = \frac{p_{t-1}}{p_t} = \rho_{t-1} \Rightarrow k^*_{t-1} = k_{t-1}\left(\rho_{t-1}\right), \tag{3.28}$$

production then equals

$$y_t(\rho_{t-1}) = f(k_{t-1}(\rho_{t-1}), 1) + (1-\delta) k_{t-1}(\rho_{t-1}), \qquad (3.29)$$

and due to constant returns to scale the real wage income is

$$\omega(\rho_{t-1}) = y(\rho_{t-1}) - \rho_{t-1} k(\rho_{t-1}).$$

The consumers are assumed to have one unit of labour when they are old and no units of labour when they are young. They therefore solve the following problem

$$\max_{c_0, c_1} U(c_0, c_1) \quad \text{s.t. } p_t c_0 + p_{t+1} c_{t+1} \le q_{t+1}. \qquad (3.30)$$

This gives consumption demand functions $c_0(\rho_t)$ and $c_1(\rho_t)$. The equilibrium condition for period t then is

$$c_0(\rho_t) + c_1(\rho_{t-1}) + k_t(\rho_t) = y_t(\rho_{t-1}), \qquad (3.31)$$

which is an implicitly defined one-dimensional map $\rho_t = F(\rho_{t-1})$. If we can establish that this map has a period three cycle then it has cycles of all periods.

We study the following specification

$$\begin{aligned} f(k_{t-1}, l_t) &= (k_{t-1}^{\sigma} + l_t^{\sigma})^{\frac{1}{\sigma}} + (1-\delta) k_{t-1}, \\ U(c_0, c_1) &= c_0^{\alpha} c_1^{1-\alpha}. \end{aligned}$$

$$U(c_0, c_1) = c_0^{\alpha} c_1^{1-\alpha}.$$

To investigate the possibility of cycles we study the corresponding general equilibrium model. This general equilibrium model has two firms and two consumers. Firm i produces good i using labour of consumer j and capital consisting of good j, $i \neq j, i, j = 0, 1$. We study the case with $\delta = \frac{1}{10}$ and $\alpha = \frac{1}{26}$. The 2×2 general equilibrium model has multiple equilibria for $\sigma < \sigma^* \approx -24.19$, so at $\sigma^* \approx -24.19$ the overlapping generations model undergoes a period doubling bifurcation. For example, for $\sigma = -25$ we have the following asymmetric equilibrium $\mathbf{p}^* = (1.0173138, 0.9826862)$ which corresponds to the following two-cycle in interest factors $\boldsymbol{\rho}^* = (0.9659617, 1.0352377)$. We also can study the extended model where there are three consumers and three firms. In that case asymmetric equilibria appear at $\sigma^c \approx -44.22$. For example for $\sigma =$

-45 we have $\mathbf{p}^* \approx (1.0921670, 0.9037847, 1.0040483)$ corresponding to a three cycle in interest factors $\boldsymbol{\rho}^* \approx (1.2084371, 0.9001407, 0.9193176)$. Therefore, given the continuity of the system, we can apply Sarkovskii's theorem and find cycles of any period for $\sigma \leq \sigma^c$. For example, for $\sigma = -35$ we have a 4-cycle $\mathbf{p}^* \approx (1.0335635, 0.9792206, 1.0494497, 0.9377662)$ or equivalently $\boldsymbol{\rho}^* \approx$(1.0554961, 0.9330801, 1.1190952, 0.9073136). Notice that for very low values of the elasticity of substitution the CES production function is in some sense close to the Leontief production function, which is the limit case for $\sigma \rightarrow -\infty$. In this limit case also 3−cycles appear.

Note that the present example corresponds to the classical case for the overlapping generations model, because wage income is obtained in the second period. The three commodities general equilibrium model, corresponding to labour supplied in the first period, gives similar results, including asymmetric equilibrium price vectors. But the difference equation of the overlapping generations model for that case is two dimensional, hence Sarkovskii's theorem cannot be applied. However, de Vilder (1995,1996) has shown that chaotic dynamics can occur in this two-dimensional version of the model, when there is a Leontief production technology and the utility functions are of the constant relative risk aversion type, similar to the ones studied by Grandmont (1985).

Chapter 4

Learning in Overlapping Generations Models

4.1 Perfect foresight versus learning

In Chapter 3 we analyzed a number of examples of perfect foresight cycles in overlapping generations models. An important feature of these perfect foresight cycles is that they are equilibrium phenomena: markets clear in every period, agents maximize their utility given their perceived budget constraints and these perceptions are correct in the sense that agents' expectations with regard to prices equal the realized values of these prices. Especially the assumption of perfect foresight may not be so innocuous as it seems at first sight. It implies that agents have a correctly specified model of the economy and are able to compute correct expectations using this model. To see that it really requires *unbounded* rationality note that an agent with perfect foresight has to know, among other things, the expectations held by all other agents. Recently the shortcomings of the perfect foresight or rational expectations approach have led an increasing number of people to study models of learning (for nice reviews on bounded rationality see Sargent (1993,1998)). The overlapping generations model with two generations has played a prominent role in this recent literature. There seem to be two reasons for this. First, the overlapping generations model provides a tractable dynamic general equilibrium model in which all agents are utility maximizers (or, given their perceptions, believe they are) and markets clear in every period. Therefore it has become a modelling tool often utilized by proponents of the rational expectations approach, especially when it comes to problems in the field of monetary economics. Secondly, under rational expectations or perfect foresight the overlapping generations model features indeterminacy: there is a continuum of perfect foresight orbits, of which all but one have the unsatisfactory characteristic that they converge to the autarkic steady state in which money has no value.[1] Indeed, some of the first models of learning in overlapping generations models were developed to provide a selection device in picking the "most likely" of the different perfect foresight paths.

In this chapter we will study a number of learning models in a standard version of the overlapping generations model. Our main focus is not on learning as a selection device but on the dynamics of different learning models itself. We belief the analysis of learning dynamics is an important subject since it may provide us with a better description of human behaviour than the rational expectations hypothesis does and it may account for some of the dynamical phenomena observed in reality. The actual dynamics of our model are given by a complicated nonlinear map. It is assumed that the agents do not know this complicated map but that they believe that the dynamics evolves according to some simple linear relationship. They estimate this relationship by using a least squares algorithm. The estimated model then provides agents' expectations about some variable, for example inflation rates. Since the actual realizations of this variable depend upon this expectation, the learning feeds back into the actual dynamics. These realizations of the variable again lead to an update of the estimated relationship. In this way a closed model of actual dynamics and learning dynamics arises. An important aspect of this model, in which it differs from most of the existing literature on learning in economic models, is the structural misperception of agents. The model they have of the economy is mis-specified. We believe that this is the most sensible approach to learning in economic models. The perceptions of an agent have to be much simpler than the unknown economic model itself which, among other things, depends upon the perception of this agent and the perceptions of all other agents in the economy and therefore will be very complicated. Agents will in

[1]This is true, provided that we are in the *Samuelson* case. For the *classical* case almost all perfect foresight paths tend to the monetary steady state. This classical monetary steady state, however, cannot be supported by money and therefore is seen as less important in macroeconomic analysis. Notice that these results refer to the overlapping generations model with two generations and one commodity per period. For an analysis of indeterminacy in more general overlapping generations model see Kehoe and Levine (1985).

general use simple "rules of thumb" to make predictions.

We investigate some different types of perceived laws of motion in this chapter. We find that different types of beliefs may lead to different stability properties of the actual dynamics and in fact various dynamical phenomena, such as the existence of periodic cycles, quasi periodic orbits and strange attractors, might occur. Furthermore it is important whether the perceived law of motion is put in terms of price levels or in terms of inflation rates, in particular when there is a positive net inflation rate in equilibrium. Clearly the number of possible perceived laws of motion is unbounded. We hope to provide more insight into this "wilderness of bounded rationality" by studying which of these perceived laws of motion are, in some sense, reasonable.

Our work is related to that of Hommes and Sorger (1998). They introduce the notion of consistent expectations equilibria. In their model agents use simple linear expectation rules in a nonlinear environment. A consistent expectations equilibrium then emerges when there are no structural errors in the forecasts agents make given their expectations function and the actual dynamics implied by this expectations function (that is, sample averages and sample autocorrelations are equal for the perceived and the actual models). In such an equilibrium agents cannot improve upon their forecasts in a linear statistical sense. Other related work can be found in the literature on learning in temporal equilibrium dynamics (Böhm and Wenzelburger (1999), Chatterji (1995), Duffy (1994) and Lettau and Van Zandt (1995)) and in the literature on least squares learning in macroeconomic models (Bray (1982), Bray and Savin (1986) and Marcet and Sargent (1989a,b)). A number of contributions has focused on the possibility of nonconvergence of least squares learning or other learning algorithms in temporal equilibrium dynamics (Benassy and Blad (1989), Bullard (1994), Grandmont and Laroque (1991) and Schönhofer (2000)). Our approach falls within this strand of the literature. Although a popular way to model learning, the least squares learning approach is not the only learning model. For example, there has been a growing interest in the genetic algorithm as a learning model. Applications of the genetic algorithm to overlapping generation models can be found in Arifovic (1995,1996) and Bullard and Duffy (1998,1999).

This chapter is organized as follows. Section 4.2 introduces the overlapping generations model we study in this chapter and briefly discusses stability properties of the perfect foresight paths. This model is taken from Bullard (1994) (and is also studied by Schönhofer (2000)). In Section 4.3 several forms of expectation formation are discussed. These are static in the sense that they are invariant over time and are not updated as new information becomes available. In Section 4.4 we study the least squares learning algorithm as put forth by Bullard (1994) and give a critical assessment of his results. We also present a learning algorithm on inflation rates, closely related to his, with significantly different stability properties. In Section 4.5 and 4.6 we present learning algorithm on perceived laws of motions that only involve inflation rates and not price levels. We show, mainly by numerical investigation, that complicated dynamical phenomena might occur, although the learning dynamics converges. Section 4.7 summarizes.

4.2 The overlapping generations model

In this section we follow Bullard (1994) and consider a standard two period overlapping generations model where in each period a generation is born that lives for two periods. Of course this is a very simple model and it can only serve as a first approximation to more complicated models with more generations, different agents and more commodities per period. We believe, however, that this simplified model is sufficiently rich to illustrate the main features of learning in more general models. The generation born in period t solves the following problem

$$\max_{c_0,c_1} U(c_0, c_1) \quad \text{subject to} \quad p_t c_0 + p^e_{t+1} c_1 \leq p_t w_0 + p^e_{t+1} w_1,$$

where $U : I\!R^2_+ \to I\!R$ is a strictly monotone, strictly quasi concave utility function, c_0 and c_1 are consumption in the first and second period of the agent's life, w_0 and w_1 are endowments in the first and second period of the agent's life, p_t is the price in period t and p^e_{t+1} is the expected price in period $t+1$. This optimization results in a savings function $S\left(\frac{p^e_{t+1}}{p_t}\right) = w_0 - c_0\left(\frac{p^e_{t+1}}{p_t}\right)$ for the young consumer. We assume that the only means of saving is money and, at least for the price paths we will study, the savings function will always be nonnegative.[2] The demand for money in period t then is

$$\frac{M_t}{p_t} = S\left(\frac{p^e_{t+1}}{p_t}\right).$$

[2]We are therefore focusing on the Samuelson case, where agents save when young and dissave when old.

The government has a series of exogenous expenditures, possibly negative[3], which it finances by seignorage (this corresponds to a tax (or subsidy) on old agents). This results in the following rule for the growth of the money stock

$$M_t = \theta M_{t-1}.$$

Equilibrium on the money market then is given by

$$S\left(\frac{p_{t+1}^e}{p_t}\right)p_t = \theta S\left(\frac{p_t^e}{p_{t-1}}\right)p_{t-1}, \tag{4.1}$$

or in terms of gross inflation rates $\pi_t \equiv \frac{p_{t+1}}{p_t}$,

$$\pi_{t-1}S\left(\pi_t^e\right) = \theta S\left(\pi_{t-1}^e\right). \tag{4.2}$$

From this last expression it can be seen that this system can have two different types of steady state equilibria: the monetary steady state $\pi^* = \theta$ and, if it exists, the autarkic steady state $\pi^* = \pi^a$, where π^a is such that $S(\pi^a) = 0$. If endowments in both periods are positive the autarkic steady state exists and if preferences are smooth and strictly convex, the autarkic steady state is unique. We will occasionally make the following assumptions on the savings function $S(.)$

Assumptions

1) *S is twice differentiable and* $S'(.) < 0$,

2) $S(\theta) > 0$.

Assumption 1 states that savings are a decreasing function of the inflation rate. This implies that c_0 and c_1 are gross substitutes. Assumption 2 states that we are in the Samuelson case where in the monetary steady state savings are positive and people transfer income from the present to the future. Assumptions 1 and 2 together imply that $\pi^a > \theta$. For convenience we introduce the following variable

$$a(\pi) = -\pi\frac{S'(\pi)}{S(\pi)}.$$

$a(\pi)$ is the (negative of the) *inflation elasticity of savings* and plays an important role in the dynamical behaviour of the overlapping generations model. Under assumptions 1 and 2 we have $a(\theta) > 0$.

[3]Negative government expenditures in this case implies that the government has a certain endowment of the commodity, which it trades for money on the market for this commodity.

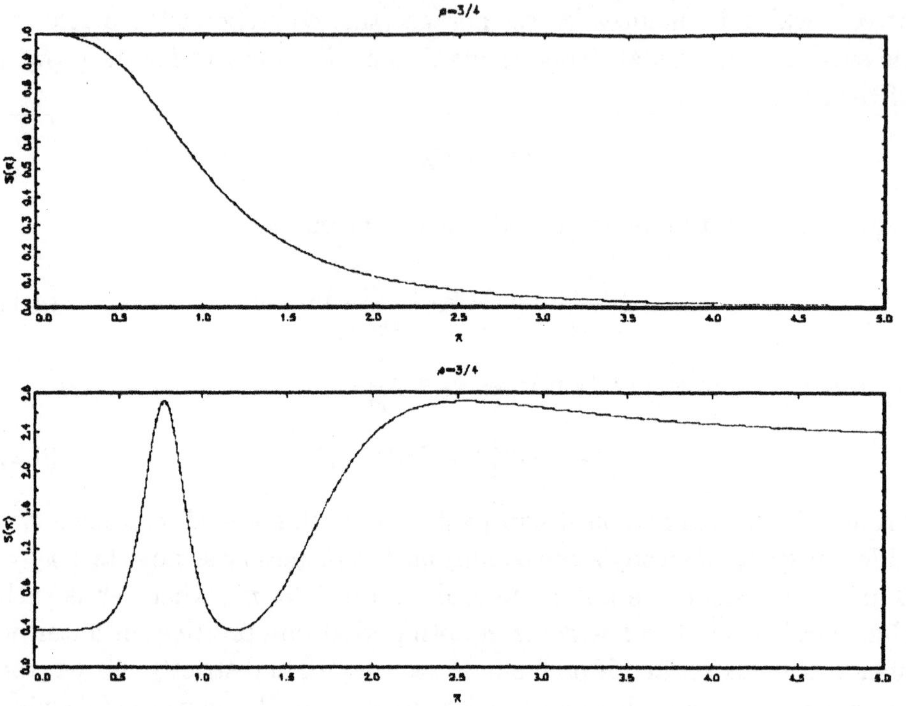

Figure 4.1: Savings functions. Upper diagram: CES savings function, lower diagram: complicated non-monotonic savings function

In order to do some numerical investigations we have to specify the savings function. Following Bullard (1994) we consider the following two typical examples. The first one derives from the well-known CES utility function $U(c_0, c_1) = (c_0^\rho + c_1^\rho)^{\frac{1}{\rho}}$ with endowments $w_0 = 1$ and $w_1 = 0$ and $\frac{1}{2} < \rho < 1$. The savings function then becomes

$$S(\pi) = \frac{1}{1 + \pi^{\frac{\rho}{1-\rho}}}.$$

Notice that it satisfies assumptions 1 and 2. Furthermore, since $w_1 = 0$ the autarkic steady state does not exist in this case. The inflation elasticity of savings for this savings function is

$$a(\pi) = \frac{\rho}{1-\rho} \frac{1}{1 + \pi^{\frac{\rho}{\rho-1}}}.$$

We will also consider a more complicated example of a savings function to illustrate some complicated phenomena that might occur in our model. The savings function is an aggregate excess demand function and therefore by the Sonnenschein-Debreu-Mantel results (see Debreu (1974), Mantel (1974,1976) and Sonnenschein (1973)) it can take on almost any form. We therefore also consider the following non-monotonic transformation of the CES savings function

$$S(\pi) = \exp\left[-\cos\left(\frac{10}{1+\pi^{\frac{\rho}{\rho-1}}}\right)\right].$$

Figure 4.1 shows these two savings functions for $\rho = \frac{3}{4}$.

Before we turn to the study of different modes of learning and expectation formation in this overlapping generations model, we briefly consider the case of perfect foresight. Under perfect foresight agents know exactly which inflation rates will obtain in the future. Therefore $\pi^e_{t+1} = \pi_{t+1}$. The overlapping generations model (4.2) then reduces to

$$S(\pi_t) = \theta \frac{S(\pi_{t-1})}{\pi_{t-1}}. \tag{4.3}$$

Under the assumption that savings decrease as the inflation rate rises, the temporal equilibrium map, implicitly defined by (4.3) is upward sloping. Therefore no complicated perfect foresight dynamics, such as cycles and chaos can occur (as we saw in Chapter 3, if the savings function is non-monotonic, cycles and even chaotic perfect foresight paths may occur). Figure 4.2 shows this temporal equilibrium map. From this figure the following well known result immediately follows.

Proposition 4.1 *Under assumptions* 1 *and* 2 *the overlapping generations model with perfect foresight (4.3) has an unstable monetary steady state* θ *and, if it exists, a locally stable autarkic steady state* π^a.

Proof. Linearizing (4.3) around a steady state π^* gives

$$S'(\pi^*)\,d\pi_t = \theta \frac{\pi^* S'(\pi^*) - S(\pi^*)}{(\pi^*)^2} d\pi_{t-1}.$$

For $\pi^* = \theta$ we have

$$d\pi_t = \left[1 - \frac{S(\theta)}{\theta S'(\theta)}\right] d\pi_{t-1} = \left[1 + \frac{1}{a(\theta)}\right] d\pi_{t-1}.$$

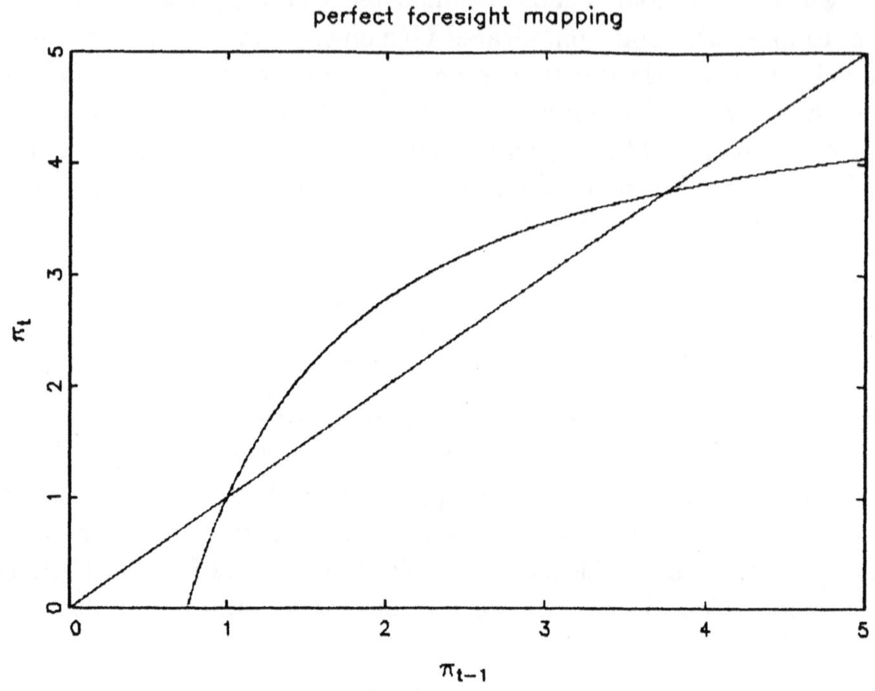

Figure 4.2: The perfect foresight mapping. The lower equilibrium corresponds to the monetary steady state θ and the higer equilibrium corresponds to the autarkic steady state π^a.

Since $a(\theta) > 0$ the monetary steady state is unstable. If an autarkic steady state exists we have for $\pi^* = \pi^a$

$$d\pi_t = \frac{\theta}{\pi^a} d\pi_{t-1},$$

and since $\pi^a > \theta$, the autarkic steady state is locally stable. ■

All orbits starting from the right of the monetary steady state converge to the autarkic steady state[4] and all orbits starting from the left of the monetary steady state become infeasible in finite time, that is the

[4]If an autarkic steady state does not exist (as in our example with CES utility functions) all orbits starting from the right of the monetary steady state then diverge to infinity.

inflation rates keep falling until for a certain π_t there is no inflation rate π_{t+1} such that the money market clears.

In the macroeconomic literature the dynamical behaviour of the overlapping generations model under perfect foresight has been perceived as disturbing for the following reason. The perfect foresight equilibrium paths are *indeterminate*, that is, there is an infinity of perfect foresight equilibrium paths, none of which is locally unique (for a discussion see Farmer (1993)). Only one of these perfect foresight paths gives the monetary steady state θ, all other perfect foresight paths converge to a steady state where money has no value. A number of contributions have focused on the multiplicity of perfect foresight paths and argued that by introducing some kind of adaptive expectations or learning the "most likely" perfect foresight path can be singled out. Lucas (1986) shows in a standard version of the overlapping generations model that if agents predict the (inverse of the) price level by using a sample average of previous observations of the (inverse of the) price level, there will be convergence to the monetary steady state. Marcet and Sargent (1989b) study the dynamical behaviour of an overlapping generations model of hyperinflation similar to ours, where agents use a least squares regression on past prices to predict future prices. They show, for their model, that there can only be convergence to the monetary steady state. They however also point out that the least squares learning algorithm may lead to instability.

It is important to note that all these results refer to the Samuelson case, where income is transferred from the young to the old. All stability results are reversed when we consider the classical case, where young agents consume more than their endowment and therefore "dissave" (that is, saving of the young is negative). Old agents then pay back the debt they incurred when they were young.

4.3 Static expectations

In the previous section we have discussed the dynamics of the overlapping generations model under perfect foresight. Clearly, the assumption of perfect foresight is a demanding one, since it requires that agents exactly know the market equilibrium equations and are able to use these to compute the market clearing prices for the next periods. From now on we will consider less divine agents, who use observations on past prices or inflation rates to make predictions about future inflation rates. In this section we will consider simple forms of backward looking expectations.

First we have to be clear about the information agents posses at the moment they have to make predictions. Agents born in period $t+1$ know all prices up to time t, and therefore the last inflation rate they have observed is $\pi_{t-1} = \frac{p_t}{p_{t-1}}$. The savings decision of these newborn only depends upon the ratio of price levels between periods $t+1$ and $t+2$. The expectation π_{t+1}^e of this price ratio is one of the determinants of the price level p_{t+1} in period $t+1$. We make the assumption that agents have already made their savings decision at the moment they enter the money market and that the prices they observe at this market therefore cannot effect this decision anymore. Hence, the expectation π_{t+1}^e and the savings decision is based upon information on inflation rates up till time $t-1$. An alternative would be to let the expectation formation of agents for π_{t+1}^e also depend upon the value of π_t. Lettau and Van Zandt (1995) show that these differences in information structure lead to significant differences in the dynamical behaviour of the overlapping generations model.[5]

We will first consider *naive* expectations. Under naive expectations agents believe that the inflation rate in the next period will be the same as in this period. Due to the information structure discussed above we then obtain the following expectations scheme: $\pi_{t+1}^e = \pi_{t-1}$. The overlapping generations model (4.2) now becomes

$$\pi_{t+1} = \theta \frac{S(\pi_{t-1})}{S(\pi_t)}. \tag{4.4}$$

A local stability analysis of the monetary steady state then reveals the following.

Proposition 4.2 *Assume that $a'(\theta) \neq 0$. Under assumptions* 1 *and* 2 *the overlapping generations model with naive expectations (4.4) generically undergoes a Hopf bifurcation at the monetary steady state when $a(\theta) = 1$. For $a(\theta) < 1$ the monetary steady state is stable and for $a(\theta) > 1$ the monetary steady state is unstable. If $a(\theta) < 0$ generically a period-doubling bifurcation occurs at the monetary steady state for that value of θ for which $a(\theta) = -\frac{1}{2}$. Then if $-\frac{1}{2} < a(\theta) < 1$ the monetary steady state is locally stable and if $a(\theta) < -\frac{1}{2}$ the monetary steady state is unstable.*

[5] Our approach is more common in the literature. One of the reasons for this is that it would complicate the analysis considerably, if we would assume that the expectation would depend upon current inflation rates. This would imply that we have to solve for an equilibrium in each period and a closed form expression for the inflation dynamics would then not be available.

Proof. The overlapping generations model can be written as a system of two first order difference equations

$$\begin{aligned} \pi_{t+1} &= \theta \frac{S(\gamma_t)}{S(\pi_t)}, \\ \gamma_{t+1} &= \pi_t. \end{aligned}$$

Linearizing this system around its steady state $(\pi^*, \gamma^*) = (\theta, \theta)$ gives

$$\begin{pmatrix} d\pi_{t+1} \\ d\gamma_{t+1} \end{pmatrix} = \begin{pmatrix} a(\theta) & -a(\theta) \\ 1 & 0 \end{pmatrix} \begin{pmatrix} d\pi_t \\ d\gamma_t \end{pmatrix}.$$

The eigenvalues of the Jacobian matrix are

$$\mu_{1,2} = \frac{1}{2} a(\theta) \pm \frac{1}{2} \sqrt{(a(\theta) - 4)\, a(\theta)}.$$

These eigenvalues are complex for $a(\theta) \in (0, 4)$. The eigenvalues cross the unit circle when

$$\mu_1 \mu_2 = a(\theta) = 1$$

Therefore a Hopf bifurcation occurs when $a(\theta) = 1$. If $a(\theta) < 0$ the eigenvalues are real. It can be easily checked that the positive root is always smaller than $+1$. The negative root equals -1 when $a(\theta) = -\frac{1}{2}$.
■

The Hopf bifurcation referred to in the proposition may be *subcritical* or *supercritical.* For a subcritical Hopf bifurcation a repelling invariant closed curve around the monetary steady state exists for $a(\theta)$ close to but smaller than 1. This closed curve coalesces with the monetary steady state at $a(\theta) = 1$. It implies that, as $a(\theta)$ approaches 1, the basin of attraction of the monetary steady state shrinks and finally disappears. For a supercritical Hopf bifurcation an attracting closed curve exists for $a(\theta)$ close to and larger than 1. On the invariant closed curve created in the Hopf bifurcation the dynamics are quasi periodic. Simulations suggest that for our overlapping generations model with CES utility functions the Hopf bifurcation is supercritical. The bifurcation value of θ can be explicitly calculated as

$$\theta^* = \left(\frac{2\rho - 1}{1 - \rho} \right)^{\frac{\rho - 1}{\rho}}, \quad \frac{1}{2} < \rho < 1. \tag{4.5}$$

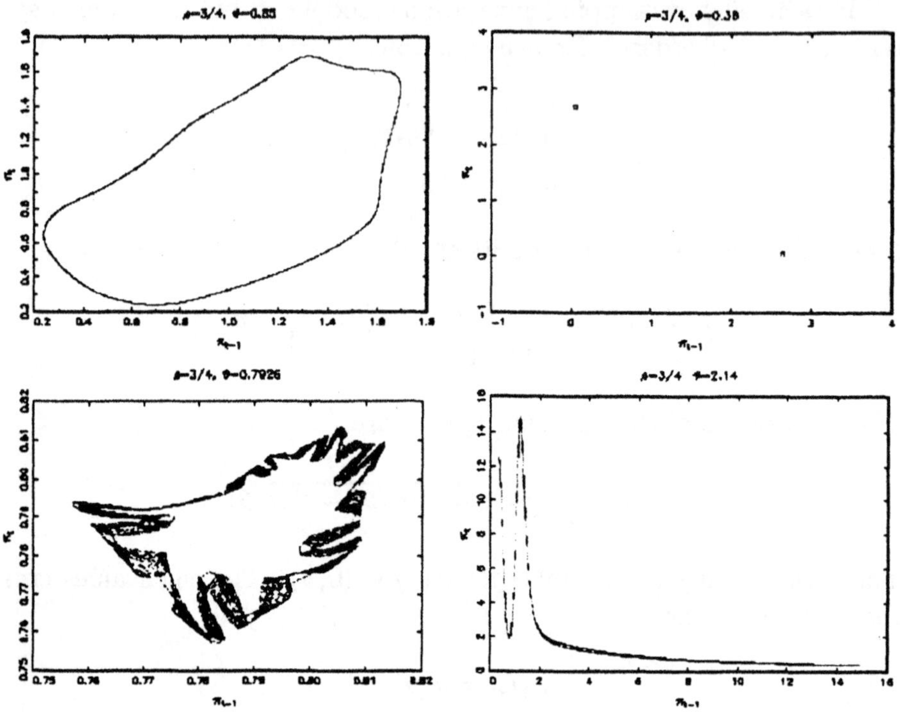

Figure 4.3: Attractors for the OG model with naive expectations. a) Invariant circle created through Hopf bifurcation for OG with CES savings function, $\rho = \frac{3}{4}$ and $\theta = 0.85$. b) Period two cycle created through flip bifurcation for the complicated savings function, $\rho = \frac{3}{4}$ and $\theta = 0.36$. c) and d) Strange attractors for OG with complicated savings function, $\rho = \frac{3}{4}$ and $\theta = 0.7926$ and $\theta = 2.14$, respectively.

Figure 4.3.a) shows an example of the attracting closed curve created through this supercritical Hopf bifurcation, for the model with CES utility functions and $\rho = \frac{3}{4}$, $\theta = 0.85 > \sqrt[3]{\frac{1}{2}} = \theta^*$.[6] The other pictures in Figure 4.3 are attractors for the overlapping generations model with the more complicated savings function. As can be seen from Figure 4.1.b) the slope of this savings function becomes positive at a certain time and

[6]This corresponds to a situation of deflation where the government also supplies the commodity and extracts a certain amount of money from the economy. For other specifications of the CES utility functions (in particular for $\rho < \frac{2}{3}$) periodic behaviour only occurs if $\theta > 1$ and there is positive inflation. The qualitative results remain the same however.

therefore $a(\theta)$ becomes negative. For $\rho = \frac{3}{4}$ and $\theta \approx 0.3535$, $a(\theta) = -\frac{1}{2}$ and a period doubling bifurcation occurs resulting in a period two orbit. This attracting period two orbit is shown in Figure 4.3b), for $\theta = 0.36$. For this complicated savings function, also some strange attractors might emerge. As can be seen from Figure 4.1.b) as θ increases $S'(\theta)$ becomes negative, hence $a(\theta)$ increases again and becomes positive. In fact, for $\theta \approx 0.7890$, the inflation elasticity of savings equals 1 and a Hopf bifurcation occurs. This Hopf bifurcation results in an invariant circle. This invariant circle undergoes a bifurcation route to a strange attractor as θ increases even more. This strange attractor is shown for $\theta = 0.7926$ in Figure 4.3.c). It coexists with the period two orbit created in the primary bifurcation. The strange attractor soon becomes unstable however, and then almost all orbits converge to the period two cycle again. As θ increases the monetary steady state becomes stable again and loses stability through a period doubling bifurcation again (compare Figure 4.1.b)). Figure 4.3.d) shows another strange attractor, for $\theta = 2.14$. This attractor is created through a cascade of period doubling bifurcations. As θ increases even more the period two orbit becomes the unique attractor once again. It disappears through a final period doubling (or period halving) bifurcation after which the monetary steady state is stable again. That this will happen can also be seen from Figure 4.1.b).

Under naive expectations agents believe that the inflation rate in the current period will be the same as in the inflation rate in the previous period. Clearly for the attractors in Figure 4.3 the agents make forecast errors. If there is no structure in these forecast errors and they look like white noise, the agents might attribute them to some stochastic noise in the economy, although in reality they might be generated by a deterministic nonlinear system. For the attractors in Figure 4.3, however, there appears to be some structure in the forecast errors.[7]

Another form of static expectations are given by so-called *adaptive expectations*. These are of the following form

$$\pi_{t+1}^e = \alpha\pi_{t-1} + (1-\alpha)\,\pi_t^e, \quad 0 < \alpha \leq 1. \tag{4.6}$$

This can also be written as

$$\pi_{t+1}^e = \pi_t^e + \alpha\left(\pi_{t-1} - \pi_t^e\right),$$

[7] Notice that due to the information structure in the overlapping generations model, it could be argued that there is no forecasting error under naive expectations if the dynamical system converges to a period two orbit. In fact, if agents would expect a period two cycle they would make the same forecasts as when they have naive expectations, but they would always be correct.

which shows the adaptive character of this type of expectations formation: expectations are updated as new information becomes available. The parameter α gives the importance that the agents attach to new observations. As is well-known we can, by repeated substitution, write the adaptive expectations scheme as

$$\pi_{t+1}^e = \sum_{j=1}^{t} \alpha (1-\alpha)^{j-1} \pi_{t-j} + (1-\alpha)^t \pi_1^e.$$

Therefore expectations depend upon all previous realizations albeit with decreasing weights. It is also clear that the influence of the initial expectation, π_1^e, decreases as time goes by. The adaptive expectations scheme is a typical example of a so-called *constant gains* learning algorithm, each new observation is as important as previous observations were (the gain of each new observation is the same as the gain from previous observations). This implies that observations in the past become increasingly less important as the above formula shows.

We can again write the overlapping generations model (4.2) with adaptive expectations (4.6) as a system of two first order difference equations in the following way

$$\begin{aligned} \pi_{t+1}^e &= (1-\alpha)\pi_t^e + \alpha\theta\frac{S(\gamma_t)}{S(\pi_t^e)} \\ \gamma_{t+1} &= \pi_t^e \end{aligned} \tag{4.7}$$

Linearizing (4.7) around the monetary steady state gives the following Jacobian matrix

$$\mathbf{J}_{adap} = \begin{pmatrix} (1-\alpha)+\alpha a(\theta) & -\alpha a(\theta) \\ 1 & 0 \end{pmatrix}. \tag{4.8}$$

The eigenvalues of this matrix are complex if $1+\alpha-2\sqrt{\alpha} < \alpha a(\theta) < 1+\alpha+2\sqrt{\alpha}$. A Hopf bifurcation then generically occurs when the determinant of the Jacobian matrix equals 1, that is, when $\alpha a(\theta) = 1$. In all other cases we have real eigenvalues and a period doubling bifurcation occurs when $a(\theta) = \frac{1}{2} - \frac{1}{\alpha}$. The local dynamics around the steady state therefore are qualitatively equivalent with those of (4.4), except that the longer memory, represented by $\alpha > 0$, stabilizes the dynamics.

The dynamical phenomena that were encountered in the study of the overlapping generations model with naive expectations, such as periodic orbits, aperiodic orbits and strange attractors can also be found in model (4.7). Since this is not our main concern at this point we will not pursue this issue any further here.

4.4 A regression on price levels

In the previous section we saw that when agents employ static expectations in order to make their forecasts for future inflation rates may result in a stable equilibrium, periodic behaviour, quasi periodic behaviour and even strange and erratic fluctuations. In this section we will describe and discuss a learning model for the overlapping generations model from the previous sections, which was developed by Bullard (1994). He assumes that agents believe that prices evolve according to the following perceived law of motion

$$p_t = \beta p_{t-1}. \tag{4.9}$$

Notice that this implies that they believe that the inflation rate is constant and equal to β. However, β is unknown to the agents and it is estimated by a regression of the price level on lagged values of the price level. That is, β is estimated by an ordinary least squares regression, using information up to time $t-1$, which gives

$$\beta_t = \frac{\sum_{s=1}^{t-1} p_{s-1} p_s}{\sum_{s=1}^{t-1} p_{s-1}^2}, \tag{4.10}$$

as the estimate of agents born in period t. Their forecast of the inflation rate then becomes $\pi_t^e = \beta_t$. Given this, the price dynamics of the overlapping generations model becomes

$$p_t = \theta \frac{S(\beta_{t-1})}{S(\beta_t)} p_{t-1}. \tag{4.11}$$

Clearly (4.10) and (4.11) together form an *expectations feedback system.* Realized prices influence perceptions agents have about economic reality and these perceptions feed back into the actual dynamics and determine which prices will be realized. We can make a distinction between two types of dynamics: the learning dynamics (4.10) which determine how agents beliefs about the economy are adapted as they receive information about the price levels and the dynamics of price levels or inflation rates (4.11) that determine how the actual price levels evolve given the beliefs of the agents.

We can write the complete system (4.10-4.11) as a recursive dynamical system by introducing the variable $g_t = p_{t-1}^2 \left[\sum_{s=1}^{t} p_{s-1}^2\right]^{-1}$. Using

(4.11) the least squares estimate at time $t+1$ can be written as

$$\begin{aligned}\beta_{t+1} &= \frac{\sum_{s=1}^{t-1} p_{s-1}p_s + p_{t-1}p_t}{\sum_{s=1}^{t} p_{s-1}^2} = \frac{\beta_t\left(\sum_{s=1}^{t} p_{s-1}^2 - p_{t-1}^2\right) + p_{t-1}p_t}{\sum_{s=1}^{t} p_{s-1}^2} \\ &= \beta_t + \frac{p_{t-1}^2\left(\frac{p_t}{p_{t-1}} - \beta_t\right)}{\sum_{s=1}^{t} p_{s-1}^2} = \beta_t + g_t\left[\theta\frac{S(\beta_{t-1})}{S(\beta_t)} - \beta_t\right].\end{aligned}$$

Furthermore we have

$$\begin{aligned}g_{t+1} &= \frac{p_t^2}{\sum_{s=1}^{t} p_{s-1}^2 + p_t^2} = \frac{p_t^2}{p_{t-1}^2 g_t^{-1} + p_t^2} \\ &= \left[\left(\frac{p_{t-1}}{p_t}\right)^2 g_t^{-1} + 1\right]^{-1} = \left[\left(\theta\frac{S(\beta_{t-1})}{S(\beta_t)}\right)^{-2} g_t^{-1} + 1\right]^{-1}.\end{aligned}$$

The whole model can then be written as a three-dimensional system of first order difference equations as follows

$$\begin{aligned}\beta_{t+1} &= \beta_t + g_t\left[\theta\frac{S(\gamma_t)}{S(\beta_t)} - \beta_t\right], \\ \gamma_{t+1} &= \beta_t, \\ g_{t+1} &= \left[g_t^{-1}\left(\theta\frac{S(\gamma_t)}{S(\beta_t)}\right)^{-2} + 1\right]^{-1}.\end{aligned} \tag{4.12}$$

Notice that the dynamics of the beliefs (β_t) coincides with the dynamics of the inflation rates, implying that if the least squares estimator β_t converges then so do inflation rates. The system (4.12) has steady state $(\beta^*, \gamma^*, g^*) = (\theta, \theta, 1 - \theta^{-2})$.

Proposition 4.3 *(**Bullard (1994)**) Assume assumptions* 1 *and* 2 *are satisfied and that* $\theta > 1$. *Then the monetary steady state generically undergoes a Hopf bifurcation at that value* θ^* *of* θ, *for which*

$$\left(1 - \theta^{-2}\right) a(\theta) = 1,$$

for $\theta < \theta^*$, *the monetary steady state is stable and for* $\theta > \theta^*$ *it is unstable.*

Proof. The Jacobian matrix of (4.12) evaluated at the equilibrium $(\theta, \theta, 1-\theta^{-2})$ equals

$$\begin{pmatrix} \theta^{-2}+\left(1-\theta^{-2}\right)a(\theta) & -\left(1-\theta^{-2}\right)a(\theta) & 0 \\ 1 & 0 & 0 \\ \frac{\partial g}{\partial \beta} & \frac{\partial g}{\partial \gamma} & \theta^{-2} \end{pmatrix}. \tag{4.13}$$

One of the eigenvalues is equal to θ^{-2} and hence stable for $\theta > 1$. The eigenvalues of the upper 2×2 matrix are complex and lie on the unit circle when $\left(1-\theta^{-2}\right)a(\theta) = 1$. ■

For our CES example with $\rho = \frac{3}{4}$ we have $\theta^* = \frac{1}{2} + \frac{1}{2}\sqrt{3} \approx 1.3660$. Two attractors for the CES savings function are shown in the upper pictures in Figure 4.4. The first picture shows the attracting invariant closed curve created in the Hopf bifurcation, the second picture shows a strange attractor that is created from this invariant closed curve. The lower pictures in Figure 4.4 show strange attractors that emerge from system (4.12) with the complicated savings function. For this complicated savings function (or in fact for any savings function with $a(\theta) < 0$) it can be shown that (4.12) undergoes a period doubling bifurcation at the monetary steady state at that value of θ for which $\frac{\theta^2-1}{\theta^2+1}a(\theta) = -\frac{1}{2}$.

Previous work on least squares learning in economic models mainly deals with establishing convergence of the beliefs to some finite limit. We now want to develop an intuition why this does not happen in the current model. Ordinary least squares algorithms are so-called *decreasing gains* algorithms. Different observations receive the same weights in the regression which implies that, as time goes by, the impact of new observations (that is the gain of new observations) becomes smaller. In (4.12) this gain is represented by the variable $g_t = p_{t-1}^2 / \sum_{s=1}^t p_{s-1}^2$. Now if price levels are bounded then, maybe except for some very special time paths, g_t will converge to 0. This would result in convergence of the inflation rate. In the model at hand, however, price levels are not bounded and in fact, in equilibrium they grow at a constant rate $\theta > 1$. This implies that the weight g_t does not converge to 0 but to a positive number $1 - \frac{1}{\theta^2}$, provided the monetary steady state is stable. Therefore, even for very large t new observations may lead to a significant change in the beliefs of the agents. In fact, if we put in the equilibrium value

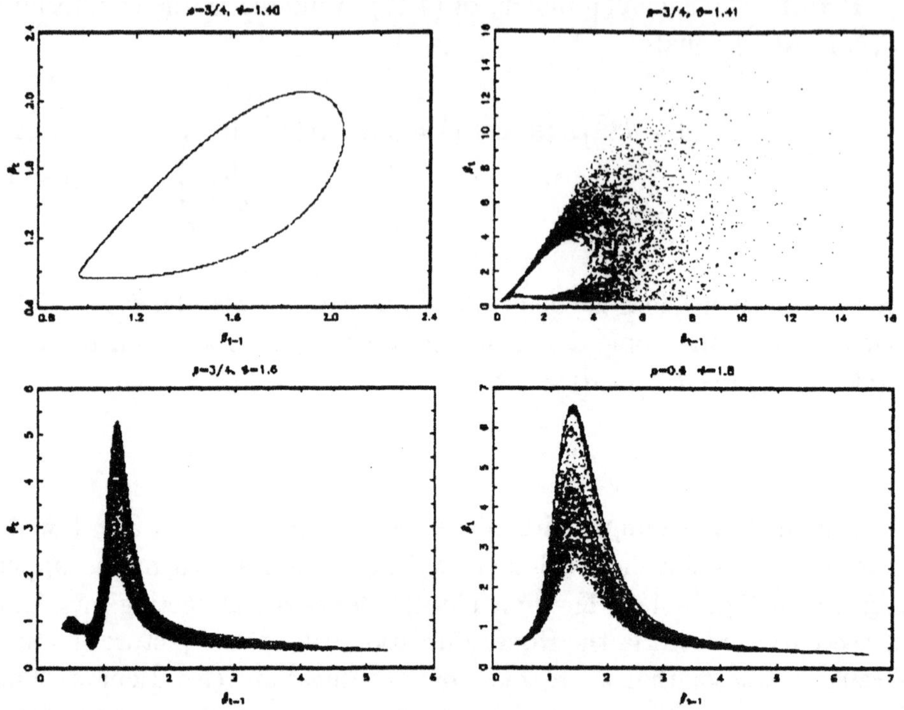

Figure 4.4: Attractors for the OG model with least squares learning on price levels. a) CES savings function, $\rho = \frac{3}{4}$ and $\theta = 1.40$. b) CES savings function, $\rho = \frac{3}{4}$ and $\theta = 1.41$. c) Complicated savings function, $\rho = \frac{3}{4}$ and $\theta = 1.5$. d) Complicated savings function, $\rho = \frac{3}{5}$ and $\theta = 1.8$.

for g_t, (4.12) reduces to

$$\begin{aligned} \beta_{t+1} &= \frac{1}{\theta^2}\beta_t + \left(1 - \frac{1}{\theta^2}\right)\theta\frac{S(\gamma_t)}{S(\beta_t)}, \\ \gamma_{t+1} &= \beta_t, \end{aligned} \tag{4.14}$$

which is just the overlapping generations model under adaptive expectations, where the weight α equals $1 - \frac{1}{\theta^2}$. The stability condition for (4.12) is exactly the same as the stability condition for the adaptive case with $\alpha = 1 - \frac{1}{\theta^2}$. This follows from the fact that the upper 2×2 matrix of (4.13) (which is the relevant part) is equal to the Jacobian of the overlapping generations model with adaptive expectations (4.8). Therefore the local stability properties of (4.12) are equivalent with those of (4.7). This explains why it can be possible that the learning and inflation dynamics do not converge.[8] Notice that the belief parameters in this case keep fluctuating.

Due to the nonstationarity of the time series of the price level an alternative approach, and a more sensible one from an econometrician's viewpoint, would be to forecast inflation rates by using the average of previous inflation rates. The agents in the previous model believe that the inflation rate is constant, and they estimate this constant by the ordinary least squares estimate of a regression of the price level on itself. The following question now arises: why don't agents estimate the inflation rate by computing the average of past inflation rates? This estimate would then be

$$\beta_t = \frac{1}{t-1}\sum_{s=0}^{t-2}\pi_s,$$

which is also the least squares estimate one obtains when regressing the inflation rate on a constant. β_t can be written recursively as

$$\beta_t = \frac{1}{t-1}\left[\sum_{s=0}^{t-3}\pi_s + \pi_{t-2}\right] = \frac{t-2}{t-1}\beta_{t-1} + \frac{1}{t-1}\pi_{t-2}.$$

The dynamical system becomes

$$\beta_{t+1} = \frac{t-1}{t}\beta_t + \frac{1}{t}\theta\frac{S(\beta_{t-1})}{S(\beta_t)}. \tag{4.15}$$

Now compare this with system (4.14). It is immediately clear that the stability properties of this system are different. Taking averages corresponds to a decreasing gains algorithm as can be seen by rewriting (4.15) as

$$\beta_{t+1} = \beta_t + \frac{1}{t}\left(\theta\frac{S(\beta_{t-1})}{S(\beta_t)} - \beta_t\right) = \beta_t + \frac{1}{t}\left(\pi_{t-1} - \beta_t\right).$$

It is clear from this formulation that, as time goes to infinity, the contribution of new observations will decrease to 0. Therefore it seems reasonable to assume that inflation rates and beliefs will converge to the

[8] The above results only hold if there is a positive amount of money creation, $\theta > 1$. If $\theta < 1$, the learning dynamics breaks down. The equilibrium value for g then becomes negative, whereas g is a ratio of strictly positive quantities, so clearly this equilibrium cannot be stable. Simulations show that in this case g approaches 0 very fast and this implies that beliefs are updated at a very low level and will not reach their equilibrium level. Inflation rates, on the other hand, do converge.

monetary steady state. Simulations support this conjecture, though convergence might be rather slow (which is also due to the decreasing gains character of (4.15)). It is however well known that time averages need not converge. Situations might arise where beliefs and inflation rates keep on changing, albeit at an increasingly slower rate. This might also happen in this case. For example, when $S(\pi)$ gets arbitrarily close to 0 for some values of π, there might be no convergence at all, although it looks like it is due to the fact that the magnitude of the adjustment becomes negligible. In the next section we will encounter an example with the same property. Other examples of nonconvergence of time averages can be found in Shapley (1964) and Gaunersdorfer and Hofbauer (1995).

To summarize, it seems that the way in which the perceived law of motion is put is an important determinant of the stability of the inflation dynamics. In order to make forecasts about future inflation rates a reasonable alternative to considering the nonstationary series of old prices seems to be to consider old inflation rates.

4.5 A regression on inflation rates – part I

Above we saw that regressing prices on past prices might induce fluctuating inflation rates and fluctuating beliefs if the money growth rate is high enough. An intuition was provided for this result and it was argued that the same belief could also be learned in a different way that would almost always lead to convergence of beliefs. In this section and in Section 4.6 we consider a different kind of learning dynamics, namely we let agents regress inflation rates on past inflation rates. In Section 4.6 we consider a perceived law of motion of the form $\pi_t = \alpha + \beta\pi_{t-1}$, whereas in the present section we assume that agents belief that inflation rates evolve according to the following relationship

$$\pi_t = \beta\pi_{t-1}. \tag{4.16}$$

Given the estimate of β_{t+1} of the generation born in period $t+1$, their forecast of the inflation rate becomes $\pi^e_{t+1} = \beta^2_{t+1}\pi_{t-1}$. The estimate β_{t+1} is obtained by a least squares regression on inflation rates, which is

$$\beta_{t+1} = \frac{\sum_{s=1}^{t-1} \pi_{s-1}\pi_s}{\sum_{s=1}^{t-1} \pi^2_{s-1}}.$$

Again it is important to be clear about the timing of the process. Agents born in the beginning of period $t+1$ do not know p_{t+1} yet, so the last inflation rate they have observed is $\pi_{t-1} = \frac{p_t}{p_{t-1}}$. They therefore make their estimates on the basis of all inflation rates up to π_{t-1}.

The estimate β_{t+1} can be written recursively again. First define $R_{t+1} = \left[\sum_{s=1}^{t} \pi_{s-1}^2\right]^{-1}$. We then have

$$R_{t+1} = \left[\sum_{s=1}^{t-1} \pi_{s-1}^2 + \pi_{t-1}^2\right]^{-1} = \left[R_t^{-1} + \pi_{t-1}^2\right]^{-1} \tag{4.17}$$

using this we can write

$$\begin{aligned}
\beta_{t+1} &= R_t \sum_{s=1}^{t-1} \pi_{s-1}\pi_s = R_t \left[\sum_{s=1}^{t-2} \pi_{s-1}\pi_s + \pi_{t-2}\pi_{t-1}\right] \\
&= R_t \left[\beta_t R_{t-1}^{-1} + \pi_{t-2}\pi_{t-1}\right] = R_t \left[\beta_t \left(R_t^{-1} - \pi_{t-2}^2\right) + \pi_{t-2}\pi_{t-1}\right] \\
&= \beta_t + R_t \pi_{t-2} \left[\pi_{t-1} - \beta_t \pi_{t-2}\right]
\end{aligned}$$

Therefore the learning dynamics can be written recursively as

$$\begin{aligned}
\beta_{t+1} &= \beta_t + R_t \pi_{t-2} \left[\pi_{t-1} - \beta_t \pi_{t-2}\right] \\
R_{t+1} &= \left[R_t^{-1} + \pi_{t-1}^2\right]^{-1}
\end{aligned} \tag{4.18}$$

Giving these learning dynamics the actual inflation rates evolve according to

$$\pi_t = \theta \frac{S\left(\beta_t^2 \pi_{t-2}\right)}{S\left(\beta_{t+1}^2 \pi_{t-1}\right)}. \tag{4.19}$$

Given the values of β_t, R_t, π_{t-1} and π_{t-2} new values β_{t+1}, R_{t+1} and π_t are determined according to the following four dimensional system of first order difference equations.

$$\begin{aligned}
\pi_t &= \theta \frac{S\left(\beta_t^2 \gamma_{t-1}\right)}{S\left(\left(\beta_t + R_t \gamma_{t-1} \left[\pi_{t-1} - \beta_t \gamma_{t-1}\right]\right)^2 \pi_{t-1}\right)} \\
\gamma_t &= \pi_{t-1} \\
\beta_{t+1} &= \beta_t + R_t \gamma_{t-1} \left[\pi_{t-1} - \beta_t \gamma_{t-1}\right] \\
R_{t+1} &= \left[R_t^{-1} + \pi_{t-1}^2\right]^{-1}
\end{aligned} \tag{4.20}$$

Notice that the difference of timing of the learning variables β and R and the inflation variables π and γ stems from the fact that expectations for people born in period $t+1$ depend upon information about inflation rates up to time $t-1$. Of course we could formulate the model in terms of new variables $\beta_t^* \equiv \beta_{t+1}$ and $R_t^* \equiv R_{t+1}$, but this would obscure the interpretation of β_{t+1} as the belief of the generation born in period $t+1$.

In contrast to the learning model studied in the previous section in the present model convergence of beliefs does not imply convergence of the inflation rate. It could be the case that learning parameters converge but that the corresponding inflation dynamics (4.19) is unstable and generates erratic inflation dynamics. Consider the possible equilibria of (4.20). Clearly, at an equilibrium we must have $\pi = \gamma = \theta$ and $\beta = 1$. At these values of π, γ and β the value of R does not matter, but from (4.17) it follows that R_t must then approach 0. Notice however that at $R=0$ the system (4.20) is not defined. To get some insights into the dynamics of (4.20) we can look at the Jacobian evaluated at $(\pi, \gamma, \beta, R) = (\theta, \theta, 1, R)$. This Jacobian J_R is

$$\begin{pmatrix} \left[1+2R\theta^2\right] a(\theta) & -\left[1+2R\theta^2\right] a(\theta) & \theta\left[1-\theta R\right] a(\theta) & 0 \\ 1 & 0 & 0 & 0 \\ R\theta & -R\theta & 1-\theta^2 R & 0 \\ -2\theta\left[R^{-1}+\theta^2\right]^{-2} & 0 & 0 & \mu_4 \end{pmatrix},$$

First of all notice that this matrix has an eigenvalue equal to $\mu_4 \equiv \left[1+R\theta^2\right]^{-2}$ which, for all positive R, lies strictly between 0 and 1. Now taking the limit of this matrix as R approaches 0 gives

$$J_0 = \lim_{R\to 0} J_R = \begin{pmatrix} a(\theta) & -a(\theta) & \theta a(\theta) & 0 \\ 1 & 0 & 0 & 0 \\ 0 & 0 & 1 & 0 \\ 0 & 0 & 0 & 1 \end{pmatrix}. \tag{4.21}$$

The Jacobian matrix J_0 has two eigenvalues equal to $+1$. These eigenvalues correspond to the learning dynamics and it is easy to see why they equal $+1$. At a point with $\pi = \gamma = \theta$ and $\beta = 1$, we saw that there is no value of R disrupting this equilibrium. Furthermore, if the weight R would be equal to 0, and abstracting from the nonexistence problem of R_{t+1}, there would be no updating of β, and therefore any β would be neutrally stable. It is important to observe that both eigenvalues are strictly between 0 and 1 for R small but positive.[9]

[9] It might be instructive to consider a model where the weight is updated as follows

$$R_{t+1} = \left[(R_t+e)^{-1} + \pi_t^2\right]^{-1},$$

The other two eigenvalues of (4.21) correspond to the inflation dynamics of (4.20), and they are

$$\mu_{1,2} = \frac{1}{2}a(\theta) \pm \frac{1}{2}\sqrt{\left(a(\theta)^2 - 4a(\theta)\right)}.$$

Notice that these are exactly the same eigenvalues as for the case of naive expectations (they correspond to the upper 2×2 matrix of (4.21)). This is intuitively clear since, if the learning dynamics converges to $\beta = 1$, agents in fact have naive expectations. Therefore, the local stability conditions of the more sophisticated learning process (4.20) are the same as that of the simplest learning process (4.4) we studied sofar, namely $-\frac{1}{2} < a(\theta) < 1$.

We are interested in the dynamical behaviour of (4.20), in particular when the monetary steady state under naive expectations is unstable. We can make a clear distinction between learning dynamics and inflation dynamics here. Due to the decreasing gains character of the least squares learning algorithm the estimate β will change very slowly as time goes by. There can however be large changes in inflation rates from period to period. Therefore it might be worthwhile to study the *short-run dynamics* of the inflation rates for a given, fixed value of β, e.g. β^0. The short-run dynamics then are given by

$$\pi_{t+1} = \theta \frac{S\left(\left(\beta^0\right)^2 \pi_{t-1}\right)}{S\left(\left(\beta^0\right)^2 \pi_t\right)}. \tag{4.22}$$

From Section 4.3 we know that these short-run dynamics can exhibit all kinds of dynamical phenomena. In particular, the inflation rates might converge to the monetary steady state, to a periodic orbit, to an invariant closed curve or even to a strange attractor. Now suppose that the short-run dynamics (4.22) indeed converge to some attractor $S_{\beta^0} \subset \mathbb{R}^2_+$. What would then happen with the long run dynamics (4.20)? Let $\{\pi_t\}_{t=1}^{\infty}$ be an orbit generated by (4.22) and lying on S_{β^0}. Then a least squares regression can be performed on these inflation rates and this will reveal an estimate for β, e.g. β^1. In general $\beta^1 \neq \beta^0$. Therefore we expect the learning dynamics to move from β^0 to β^1. But, given β^1, the short run

where $e > 0$ is small. In this case the equilibrium weight would be $R_e = \frac{1}{2}\sqrt{e^2 + 4\frac{e}{\theta^2}} - \frac{1}{2}e > 0$ and no problems would arise at the equilibrium $(\theta, \theta, 1, R_e)$. We then get results similar as the ones in the chapter. Of course, in this case the estimate β_t will not coincide with the least squares estimates of β.

dynamics will in general not converge to the attractor S_{β^0}. Therefore the estimate of β will change again. In this way we can in fact construct an implicit map of beliefs

$$\beta^{k+1} = T\left(\beta^k\right),$$

where the mapping $T(.)$ gives the ordinary least squares estimate β for the relationship $\pi_t = \beta\pi_{t-1}$, given that the inflation rates are generated by the short-run dynamics (4.22) with fixed β^k. This artificial beliefs map can help us understand the dynamics of the long run dynamics (4.20). Our main concern are the equilibria of $T(.)$, since at such an equilibrium β^* agents believe that $\pi_t = \beta^*\pi_{t-1}$ and running a regression on inflation rates indeed does not give them reason to change their beliefs. So, although agents beliefs are mis-specified, new observations on inflation rates are consistent with these beliefs in the sense that these new observations do not modify these beliefs. This equilibrium belief is the best one in the class of beliefs that is considered by the agents. This leads us to give, loosely, the following general definition of a *beliefs-equilibrium.*

Definition 4.1 *Let G be a class of perceived laws of motion. An element g of G has the following general form*

$$\pi^e_{t+1} = g\left(\pi_{t-1}, \pi_{t-2}, \ldots, \pi_{t-l}\right) = g\left(\overrightarrow{\pi}_{t-1}\right),$$

where the functional specification of g and the number of lags l is prescribed by the class G. Consider the dynamical system

$$\pi_t = \theta \frac{S\left(g^2\left(\overrightarrow{\pi}_{t-2}\right)\right)}{S\left(g^2\left(\overrightarrow{\pi}_{t-1}\right)\right)}, \tag{4.23}$$

where $g \in G$. Then we call $g^* \in G$ a Beliefs-Equilibrium *(BE) if it is the perceived law of motion in G that fits the time series of inflation rates generated by (4.23) with g^* better*[10] *than all other elements of G.*

In the present section the class G corresponds to all perceived laws of motion of the form $\pi_t = \beta\pi_{t-1}$.[11] One example of a beliefs-equilibrium is

[10] "Better" clearly is a rather subjective notion. In this chapter, where we only consider linear perceived laws of motion, one belief is "better" than another belief if its unweighted sum of squared forecast errors is smaller. This, of course, is consistent with using the (recursive) least squares algorithm as a learning model.

[11] We already encountered another general class G where we considered perceived laws of motion of the type $\pi^e_{t+1} = \beta$. Clearly for this class the unique BE is given by $\pi^e_{t+1} = \theta$.

the situation, discussed above, where the overlapping generations model with naive expectations has a stable monetary steady state. In that case $\beta^* = 1$ and indeed one has (for t high enough) $\pi_t = \beta^* \pi_{t-1} = \pi_{t-1}$. More interesting is a BE where inflation rates do not converge to the monetary steady state. An example is shown in Figure 4.5.a). Here we have considered the overlapping generations model with learning on inflation rates (4.20) with a CES savings function with $\rho = \frac{3}{4}$ and $\theta = 1$. From Section 4.3 we know that for these values the monetary steady state is unstable under naive expectations. For our model (4.20) the beliefs then converge to $\beta^* \approx 0.97346$ and the inflation dynamics converge to an attracting invariant closed curve. It can easily be checked numerically that here β^* is indeed an equilibrium of $T(.)$. Figure 4.5.a) shows the attractor of the inflation rates (the closed curve) and the beliefs of the agents (the straight line through the origin with slope β^*). One might argue that, when the agents observe this picture they might realize that their belief is mis-specified and try to learn something different. Figure 4.5.b) shows the results for the same model, only with a small noise on the parameter of the money growth. In fact, we assume $\theta_t = \theta + \varepsilon_t$ where ε_t are IID disturbances that have a normal distribution with mean 0 and standard deviation 0.02. If agents see this diagram it might indeed be sensible to run a regression on past inflation rates.

Figure 4.6 shows the time series for π_t and β_t, corresponding to Figure 4.5.a). It is clear that the latter converges and the former keeps on fluctuating (notice that the time scales are different for these two time series).

There can also be nonexistence of an equilibrium of these learning dynamics. Examples of this nonexistence can be found for the overlapping generations model with the complicated savings function. The corresponding pictures are the lower ones in Figure 4.5. Here no equilibrium β^* of the map $T(.)$ exists. To explain what happens consider the time series for the case with $\rho = \frac{3}{4}$ and $\theta = 1$, depicted in Figure 4.7.

Suppose that initial beliefs are such that β is rather small and the corresponding short-run dynamics (4.22) are stable. In that case inflation rates converge to $\theta = 1$ and the estimate of β will increase. At a certain time this β will be so high that the short-run dynamics become unstable and a period two orbit emerges. This happens at $\beta \approx 0.5930$. Then β decreases again[12] and this stabilizes the short-run dynamics (4.22).

[12] In fact, we can explicitly compute the value of β corresponding to a period two orbit (π_1, π_2). Since we know that for a period two orbit of the overlapping generations

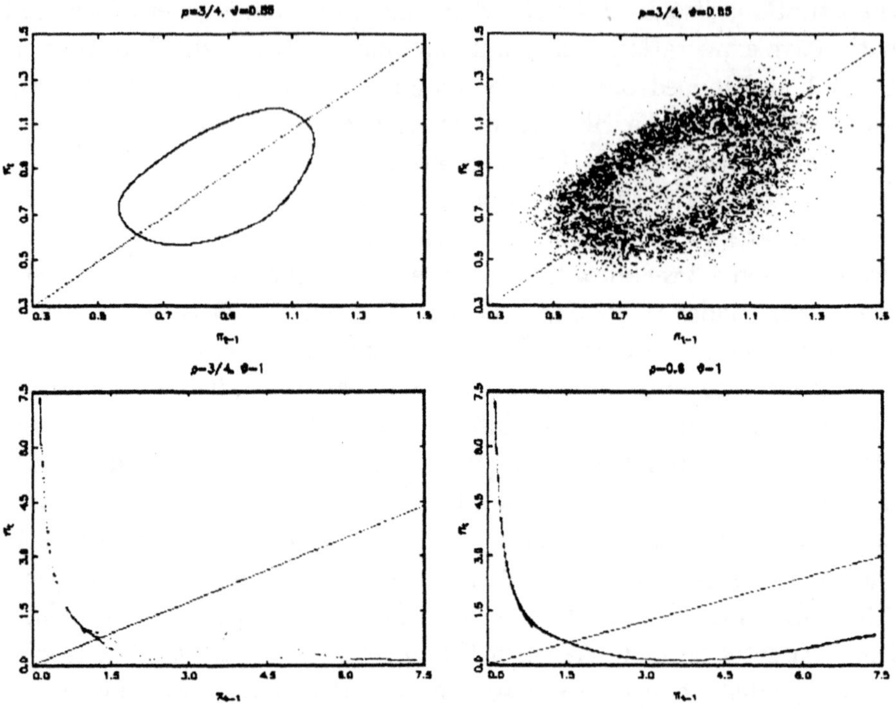

Figure 4.5: Attractors of (4.20) and examples of beliefs-equilibra. a) BE for CES savings function with $\rho = \frac{3}{4}$ and $\theta = 0.85$. b) BE from a) with small noise. c) and d) Nonexistence of BE for complicated savings function with $\rho = \frac{3}{4}$ and $\rho = \frac{3}{5}$, respectively and $\theta = 1$.

This repeats over and over. Notice that the inflation rate is stable for a long time which implies that agents will start believing more and more that it will be the same as in the previous period (which seems to be a plausible expectation) and this leads the inflation to become unstable. As discussed in the previous section we here have a situation where the least squares algorithm does not converge, though it moves slower and slower as time goes by.

model one must have $\pi_1 \pi_2 = \theta^2$, the least squares regression estimate β converges to

$$\beta = \frac{\sum_t \pi_t \pi_{t-1}}{\sum_t \pi_{t-1}^2} = \frac{2\theta^2 \pi_1^2}{\theta^4 + \pi_1^4},$$

which is always smaller than 1 for $\pi_1 \neq \theta$.

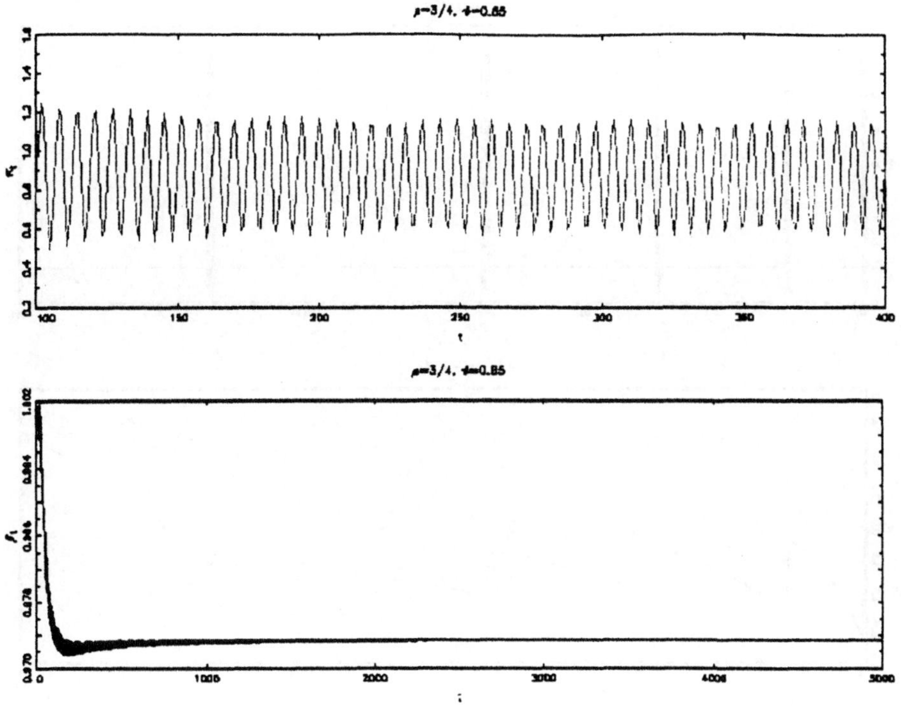

Figure 4.6: Time series of π_t and β_t for (4.20) with CES savings function and $\rho = \frac{3}{4}$ and $\theta = 0.85$.

4.6 A regression on inflation rates – part II

In the previous section we saw that regressing inflation rates on past inflation rates may lead to a situation where the belief parameters, which are "slow" variables, converge, whereas inflation rates keep on fluctuating. It can be argued that the class of perceived laws of motion we considered there is too simple. Unless $\beta = 1$, agents believe that the inflation rate will converge to zero or diverge to infinity, which clearly is not the case. In particular, sample averages are inconsistent with the perceived law of motion. In this section we want to consider a more general class of perceived laws of motion, namely

$$\pi_t = \alpha + \beta\left(\pi_{t-1} - \alpha\right). \tag{4.24}$$

Notice that $\alpha = 0$ corresponds to the perceived law of motion from the previous section and $\beta = 1$ corresponds to naive expectations. For general α and β, an agent beliefs that the long run equilibrium equals α and

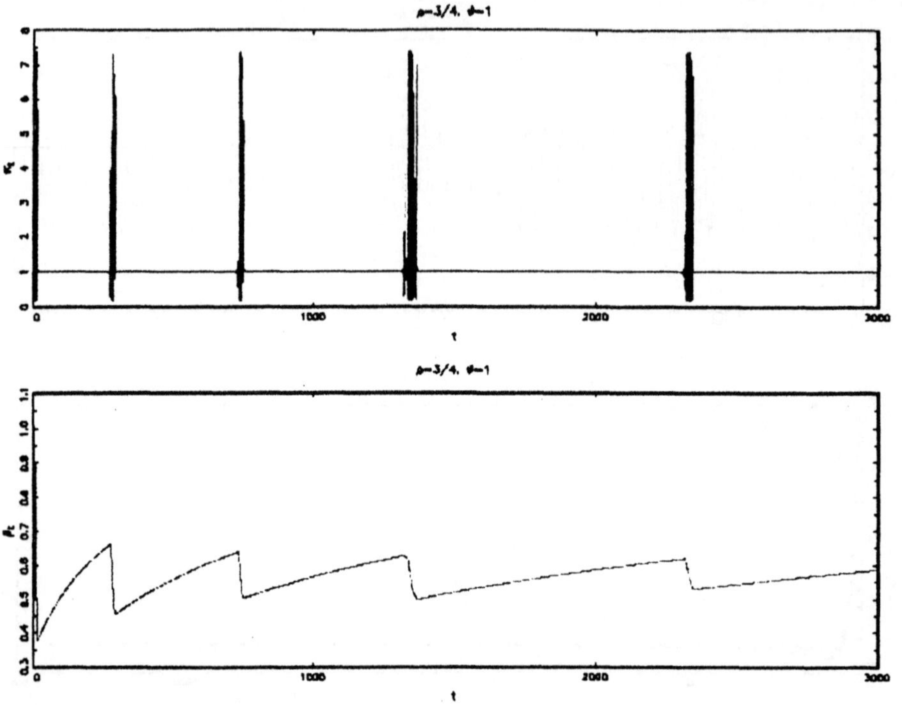

Figure 4.7: Time series of π_t and β_t for (4.20) with complicated savings function and $\rho = \frac{3}{4}$ and $\theta = 1$.

he beliefs the inflation rate will converge to this long run equilibrium at a rate β (provided that $|\beta| < 1$). If $\beta < 0$ he expects some overshooting, if $\beta > 0$ he expects monotonic convergence. The two periods ahead prediction with (4.24) becomes

$$\pi_{t+1}^e = \alpha + \beta^2 (\pi_{t-1} - \alpha)$$

and given these expectations the overlapping generations model (4.2) becomes

$$\pi_{t+1} = \theta \frac{S\left(\alpha + \beta^2 (\pi_{t-1} - \alpha)\right)}{S\left(\alpha + \beta^2 (\pi_t - \alpha)\right)}. \tag{4.25}$$

In the previous section we saw that if expectations would be correct in equilibrium, which seems to be a minimal consistency condition to be imposed on the expectations scheme, one needs $\beta = 1$. For expectations schemes of the form (4.24) to be consistent in equilibrium we need

$$\pi_t^e = \alpha + \beta(\theta - \alpha) = \theta.$$

In this case the equilibrium belief parameters α and β are unidentified. That is, there are a lot of values of α and β that satisfy the above consistency condition. In particular, all pairs with $\alpha = \theta$ and all pairs with $\beta = 1$ satisfy it. This multiplicity of beliefs equilibria becomes important when we introduce learning. In particular, we know that the inflation dynamics depend upon the parameters α and β so there are always consistent expectations schemes for which the monetary steady state is locally stable.[13].

The expectations scheme (4.24) also admits the possibility of perfect prediction when the inflation rates move in a period two cycle. Let (π_1, π_2) be such a two-cycle of the dynamics. Then correct prediction requires α and β to satisfy

$$\pi_2 = \alpha + \beta(\pi_1 - \alpha) \text{ and } \pi_1 = \alpha + \beta(\pi_2 - \alpha),$$

or equivalently, the two periods ahead prediction, given that at this time the inflation rate is π_1, must be

$$\pi_1 = \alpha + \beta^2(\pi_1 - \alpha)$$

given that $\pi_1 \neq \pi_2 \neq \alpha$ this requires that $\beta^* = -1$. α then has to equal $\frac{1}{2}(\pi_1 + \pi_2)$. Of course this situation only occurs when the inflation dynamics given these values of α^* and β^* indeed has a period two-cycle. That is

$$\pi_{t+1} = \theta \frac{S\left(\alpha^* + (\beta^*)^2(\pi_{t-1} - \alpha^*)\right)}{S\left(\alpha^* + (\beta^*)^2(\pi_t - \alpha^*)\right)} = \theta \frac{S(\pi_{t-1})}{S(\pi_t)}$$

must have a period two cycle.

[13]It can easily be calculated that, for a monotone decreasing savings function, a Hopf bifurcation occurs at that value of θ^* for which we have

$$\frac{\beta^2\theta^*}{\alpha + \beta^2(\theta^* - \alpha)} a\left(\alpha + \beta^2(\theta^* - \alpha)\right) = 1,$$

which at an equilibrium where agent have consistent expectations becomes

$$\beta^2 a(\theta^*) = 1.$$

Clearly, there are then always values of β for which the inflation dynamics are stable.

Thus expectations can be correct along the steady state and along a period two cycle. These expectations correspond to the so-called *consistent expectations equilibria* (CEE) introduced by Hommes and Sorger (1998). In their framework an expectations scheme is a CEE if agents make no structural forecast errors in a linear statistical sense. Besides steady state and two-cycle CEE they also have chaotic CEE, where the dynamics of the state variable moves over a chaotic attractor but there is no structure in the forecast errors. These CEE are related to the beliefs-equilibria we defined in the previous section. We now turn to the analysis of the learning model where agents have expectations according to (4.24) and the belief parameters α and β are learned trough a least squares regression on inflation rates. We assume agents estimate the following perceived law of motion

$$\pi_t = \beta_0 + \beta_1 \pi_{t-1},$$

which is equivalent to the perceived law of motion (4.24) up to the transformation $\beta_0 = \alpha(1-\beta)$ and $\beta_1 = \beta$. Define

$$X_{t+1} = \begin{pmatrix} 1 & \pi_0 \\ 1 & \pi_1 \\ \vdots & \vdots \\ 1 & \pi_{t-2} \end{pmatrix}, \; Y_{t+1} = \begin{pmatrix} \pi_1 \\ \pi_2 \\ \vdots \\ \pi_{t-1} \end{pmatrix} \text{ and } x_{t+1} = \begin{pmatrix} 1 & \pi_{t-1} \end{pmatrix}.$$

X_{t+1} is the set of regressors for the generation that is born in period $t+1$. Recall, that the last observation on inflation rates of these agents is π_{t-1}. The estimator b for $(\beta_0, \beta_1)'$ at time $t+1$ is

$$b_{t+1} = \left(X'_{t+1}X_{t+1}\right)^{-1} X'_{t+1}Y_{t+1} \equiv R_t X'_{t+1} Y_{t+1}$$

As before we can write this estimator recursively,[14] and the learning dynamics then are given by

$$\begin{aligned} b_{t+1} &= b_t + R_t x'_{t+1}\left[\pi_{t-1} - x_{t+1}b_t\right], \\ R_{t+1} &= \left(R_t^{-1} + x'_{t+2}x_{t+2}\right)^{-1}, \end{aligned}$$

[14]We have $R_t = \left(X'_tX_t + x'_{t+1}x_{t+1}\right)^{-1} = \left(R_{t-1}^{-1} + x'_{t+1}x_{t+1}\right)^{-1}$ and

$$\begin{aligned} b_{t+1} &= R_t X'_{t+1} Y_{t+1} = R_t\left[X'_tY_t + x'_{t+1}\pi_{t-1}\right] = R_t\left[R_{t-1}^{-1}b_t + x'_{t+1}\pi_{t-1}\right] \\ &= R_t\left[\left(R_t^{-1} - x'_{t+1}x_{t+1}\right)b_t + x'_{t+1}\pi_{t-1}\right] = b_t + R_t x'_{t+1}\left[\pi_{t-1} - x_{t+1}b_t\right]. \end{aligned}$$

and the inflation dynamics are given by

$$\pi_t = \theta \frac{S\left(\alpha_t + \beta_t^2\left(\pi_{t-2} - \alpha_t\right)\right)}{S\left(\alpha_{t+1} + \beta_{t+1}^2\left(\pi_{t-1} - \alpha_{t+1}\right)\right)}.$$

In terms of β_0 and β_1 the whole system becomes

$$\begin{aligned} b_{t+1} &= b_t + R_t x'_{t+1}\left[\pi_{t-1} - x_{t+1} b_t\right] \\ R_{t+1} &= \left(R_t^{-1} + x'_{t+2} x_{t+2}\right)^{-1} \\ \pi_t &= \theta \frac{S\left(\beta_{0t}\left(1+\beta_{1t}\right) + \beta_{1t}^2 \pi_{t-2}\right)}{S\left(\beta_{0,t+1}\left(1+\beta_{1,t+1}\right) + \beta_{1,t+1}^2 \pi_{t-1}\right)} \end{aligned} \tag{4.26}$$

Notice that these dynamics constitute an 8-dimensional system, since b_t is a 2×1 vector, R_t is a 2×2 matrix and the inflation rate in time t depends upon π_{t-1} and π_{t-2}.

In the model of the previous section we saw that there was only one equilibrium of the dynamics (corresponding to $\beta = 1$) which implies that the stability of this equilibrium is equivalent with the stability properties of the overlapping generations equilibrium under naive expectations. In the model studied in this section the situation is a little more complicated. Since there is a continuum of beliefs equilibria consistent with the monetary steady state and the stability of this monetary steady state depends upon the belief parameters α and β, there are always a number of stable equilibria and a number of unstable equilibria. In particular, there exists a number $\overline{\beta} > 0$, such that all equilibria with $|\beta^*| < \overline{\beta}$ are stable under the learning dynamics discussed above and all other equilibria are not. The initial conditions on inflation rates and the initial beliefs therefore seem to be quite important for the dynamics. It might be worth mentioning here that the least squares learning algorithms we have been studying can have an interpretation as a Bayesian learning scheme. In particular, the above algorithm also specifies the case where agents have a normally distributed prior on α and β and after each realization update this prior according to Bayes' rule (see Bray and Savin (1986) and Zelner (1971)). The matrix R_0 then corresponds to the precision of the initial beliefs. If R_0 is large, then agents are not to sure about their initial beliefs (their prior distribution has a relatively high variance) and if R_0 is small they have a lot of confidence in their initial beliefs. Therefore, if for example according to their prior the agents know for sure that $\beta_1 = 0$, (that is, they attach probability 1 to this event) the updating scheme discussed above is exactly equivalent to an

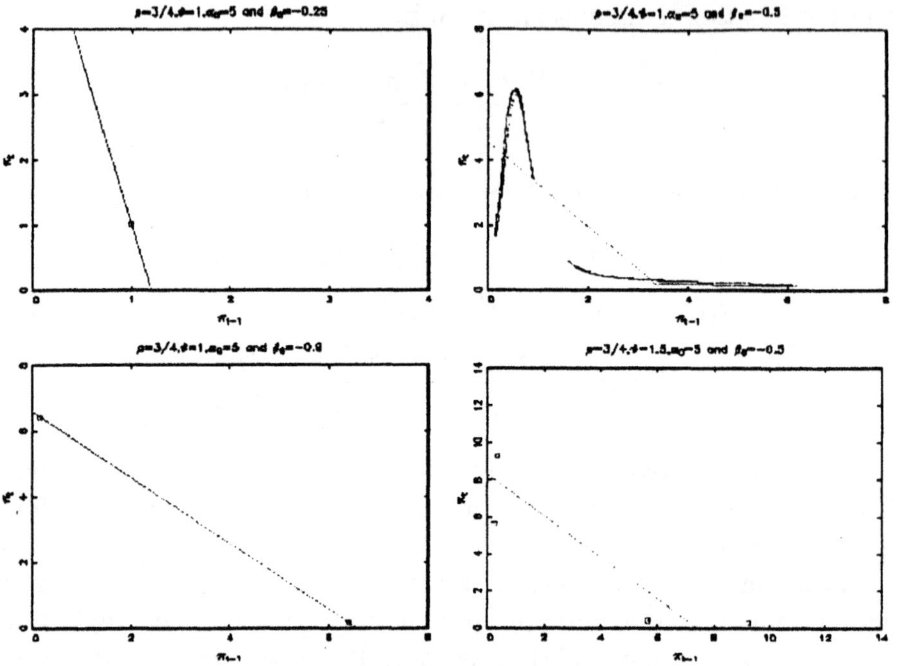

Figure 4.8: Attractors of (4.26) and examples of beliefs-equilibria with complicated savings function. a) BE corresponding to steady state, $\rho = \frac{3}{4}$, $\theta = 1$, $\alpha_0 = 5$ and $\beta_0 = -\frac{1}{4}$. b) Nonexistence of BE, $\rho = \frac{3}{4}$, $\theta = 1$, $\alpha_0 = 5$ and $\beta_0 = -\frac{1}{2}$. c) BE corresponding to period two cycle, $\rho = \frac{3}{4}$, $\theta = 1$, $\alpha_0 = 5$ and $\beta_0 = -0.9$. d) BE corresponding to four cycle, $\rho = \frac{3}{4}$, $\theta = 1.5$, $\alpha_0 = 5$ and $\beta_0 = -\frac{1}{2}$.

updating scheme where agents forecast inflation rates by taking the average of previous inflation rates, which converges almost always as we saw before.

Simulations of the overlapping generations model with the CES savings function suggest that, for all parameter values, there will be convergence to the monetary steady state and to belief parameters α and β for which the steady state is stable and which correctly predict this steady state. Which belief parameters that will be depends upon the initial conditions. For the complicated savings function the situation is a little different. Figure 4.8 shows some typical examples of the learning and inflation dynamics for this case.

Initial beliefs appear to be very important. The first three pictures show the possible dynamics of the overlapping generations model with $\rho = \frac{3}{4}$ and $\theta = 1$. For all three pictures, the initial belief on α equals 5, they only differ in the initial belief over β. The initial inflation values are the same for each simulation and are chosen near the period two orbit (which is the attractor for the overlapping generations model under naive expectations). In the first picture $\beta_0 = -\frac{1}{4}$ and as can be seen the inflation dynamics converges. The learning dynamics also converges to a situation with $\alpha = \theta = 1$. As discussed before the final value of β is indeterminate and very much depends upon the way in which the inflation rate converges to the monetary steady state θ. The second picture shows what happens when the initial value of β is a bit lower, $\beta_0 = -\frac{1}{2}$. In this case the inflation dynamics seems to converge to a strange attractor. However, this is a case of nonexistence, similar to the one discussed in the previous section. Starting with $\alpha = 5$ and $\beta = -\frac{1}{2}$ the inflation dynamics converges to a two-cycle. This makes β converge to -1, however as β becomes approximately -0.76 the period two cycle becomes unstable and a strange attractor emerges. Across this strange attractor we have a regression coefficient β smaller than -0.76 which makes the system return to a stable period two cycle again and the whole story repeats. A brief look at the time series would then reveal the following. The economy moves in an approximate two-cycle for a long period. As agents start believing more and more in this two-cycle it is disturbed and inflation rates start fluctuating more and more causing the beliefs to move in the reverse direction which leads to the two-cycle again. The third picture shows the inflation and learning dynamics when the initial belief on β equals -0.9. The inflation rates converge to a period two orbit (π_1, π_2) and the learning dynamics converges to $(\alpha, \beta) = (\frac{\pi_1+\pi_2}{2}, -1)$. Agents really have to believe in a period two cycle in order for it to indeed emerge. The fourth picture shows the dynamics for $\theta = 1\frac{1}{2}$. We then find that for initial beliefs $\alpha = 5$ and $\beta = -\frac{1}{2}$ the inflation rates converge to a period four orbit.

4.7 Summary

In this chapter we have considered a standard overlapping generations model and discussed several ways in which agents might learn about the dynamics of inflation. We have seen that, dependent upon the specific learning procedure at hand and the specification of the overlapping generations model, the monetary steady state might be stable or unstable and periodic, quasi periodic and strange behaviour might occur. An important aspect of our models is that agents perceptions are mis-specified. It is clear that for any agent it is impossible to understand the whole economy including the specific psychology and behaviour of all agents in it. Therefore, an agent always has some extremely simplified model of the economy in his mind when he tries to make forecasts of some economic variable. Therefore the use of mis-specified models of the economy is dominant in economic life.

Some important lessons may be learned from the analysis conducted above. First of all it is clear that convergence to a steady state depends very much upon the specific type of learning models people use. Secondly, erratic dynamics can not be dismissed by a higher degree of sophistication of the learning models. Thirdly, though the learning dynamics might converge to a limit belief the corresponding inflation dynamics might be erratic. These fluctuating inflation rates are then in some sense consistent with the limit belief of the agents: in their class of models of the economy, the limit belief explains the inflation rates the best. Of course models from other classes might perform better, however the class of learning models has to be taken exogenously.

A next step in the analysis would be to consider more general classes of perceived laws of motion and investigate which of these laws of motion and corresponding beliefs-equilibria are the most relevant for the model under study. In this way one might be able to restrict the wilderness of bounded rationality by ruling out laws of motion that make structural forecast errors.

Of course the analysis in this chapter has only been a starting point in a more definite characterization of the dynamic properties of learning procedures. However we believe that it carries in it all the phenomena that one would find in more realistic dynamic economic models where agents can use more sophisticated learning algorithms.

Chapter 5

An Evolutionary Model of Cournot Competition

5.1 Introduction

Up till now we have discussed general equilibrium models with perfect competition. In this and the following chapter we will consider models of imperfect competition. Probably the best known model of imperfect competition is the homogeneous Cournot duopoly model. In this chapter we will introduce a boundedly rational approach to this Cournot duopoly model. In Chapter 6 we will consider models where firms produce heterogeneous commodities. There the main focus will be on price setting behaviour.

As was discussed in Chapter 4, the assumption of perfect foresight or rational expectations is a very demanding one and casual observation suggests that it is not the proper way to describe human behaviour. Knowing the correct model of the economy implies that rational agents know in which way all other agents form expectations and behave. In particular in game theoretic models, where interaction between agents plays an important role, this seems to be important, since even if the (Nash) equilibrium of a particular game is known to all agents, it does not have to be optimal for an agent to play this equilibrium if some of its opponents do not play it. We didn't touch upon the subject of heterogeneity in Chapter 4. There we analyzed learning in the context of a representative agent model, implying that all agents behave alike. In this chapter we consider heterogeneity between agents, where different agents might behave differently. We focus on an evolutionary game theoretic model of Cournot competition where in each time period the agents are matched in pairs to play a Cournot game. A standard evolutionary game theoretic model deals with a population of agents who are programmed to pure strategies in a symmetric two-player game in normal form. Extensive discussions on several evolutionary game theoretic models can be found in the books by Samuelson (1997), Vega-Redondo (1996) and Weibull (1995). In contrast to the common practice in evolutionary game theory to study games with a finite action space, we assume a continuous and therefore infinite action space. To be able to introduce some evolutionary competition we follow Brock and Hommes (1997) and let individuals choose their behavioural rule, i.e. routine, from a finite set of different rules. These rules are functions of past information and have a publicly available performance measure. Furthermore, the costs associated with a certain rule depend on the information and computational requirements of the rule. Individuals make a choice concerning the behavioural rules based upon their past performance and costs. Each behavioural rule specifies the quantity to be produced in the next period as a function of current or past production levels. Depending on profits made by such a behavioural rule, the fraction of the population using this rule will change, according to some measure of evolutionary fitness. This results in a nonlinear dynamical system with respect to the choice of behavioural rules that is coupled with the dynamics of the endogenous quantities supplied in the Cournot game.

Our concern is whether or not an individual may want to switch between behavioural rules over time. Consider the following scenario. All individuals in a population can either behave according to a sophisticated behavioural rule H_1 at small but positive costs or obtain a simple behavioural rule H_2 for free. Each time period all individuals are randomly matched in pairs to play a Cournot duopoly game. They use the behavioural rules to determine the quantity they supply in this stage game and they tend to choose the rule that yields the highest expected net profit. Suppose that if all players use the sophisticated rule H_1, all time paths of the endogenous variables, i.e. the quantities supplied in the Cournot game, would converge to a unique stable equilibrium, which is the well known Cournot-Nash equilibrium. Furthermore, assume that when all agents use the simple rule H_2, the Cournot-Nash equilibrium is unstable. If quantities start out close to the Cournot-Nash equilibrium and almost all players use the simple rule H_2, quantities will diverge from the equilibrium and average profits from using rule H_2 will decrease. As a result, the number of players willing to pay some information costs to use the sophisticated behavioural rule H_1 increases. As soon as enough agents have switched to H_1, quantities will be pushed back towards their equilibrium values. With quantities close to their equilibrium value, the average net profit of the simple rule H_2 becomes higher, while the players using the sophisticated rule H_1 still have to pay the information costs. Consequently, most agents will switch to using the simple rule H_2 again, and the story repeats. In this chapter we analyze two typical examples where the mechanism described above leads to erratic dynamics. For one of these examples we show that strange attractors exist. We do this by applying new results from *homoclinic bifurcation theory* to our evolutionary Cournot model. The use of homoclinic bifurcation theory is rather new in economic theory. It has been applied in the aforementioned work of Brock and Hommes (1997) and the work of de Vilder (1995,1996) who studies the appearance of homoclinic bifurcations in a two-dimensional overlapping generations model with capital.

Another model in which agents are identified with a behavioural rule is discussed in Axelrod (1984). Axelrod describes two computer tournaments for which he invited various people to submit automata that would then compete against each other in the repeated Prisoner's Dilemma. The tit-for-tat automaton won both times. His next step was to simulate the effect of further tournaments. He began with an initial population in which each of the automata used in his second tournament was equally numerous. He then used a version of the replicator dynamics to make a guess at the profile of automata that would have been submitted in later tournaments. In the long run, the process settled on a polymorphic profile in which tit-for-tat had the largest share. The survival of the tit-for-tat automaton shows that cooperation in the Prisoner's Dilemma is sustainable.

Related issues concerning the selection and evolution of behavioural or learning rules are discussed in Björnerstedt and Schlag (1994) and Kirchkamp and Schlag (1994), respectively. Banerjee and Weibull (1995) analyze the survival of nonrational agents in a strategic environment represented as a symmetric two-player game. Furthermore, see Vega-Redondo (1997) for an evolutionary model of Cournotian oligopoly and Rhode and Stegeman (1995) for a duopoly scenario.

This chapter is organized as follows. In Section 5.2 we introduce the general evolutionary model and the corresponding dynamical system. In addition, we state some local instability results. In Section 5.3 we introduce our leading example, where agents can choose between two different behavioural rules. In Section 5.4 we analyze the local bifurcations of the model for two types of population dynamics: the discrete choice model and the replicator dynamics. We also provide some simulation results there. In Section 5.5 we consider the global dynamics of the replicator dynamics and show that a homoclinic bifurcation occurs. Another typical example with two behavioural rules is briefly studied in Section 5.6 and Section 5.7 concludes. The Appendices contain some of the computations needed for the main results of this chapter.

5.2 The Model

Consider an infinite population of players who are matched in pairs randomly each discrete-time period to play a symmetric Cournot duopoly game. However, instead of simultaneously choosing the supplied quantities directly, the players act according to behavioural rules that exactly prescribe the quantity to be supplied. Note that even though the model is formulated for the duopoly case, it can easily be generalized to allow for any finite number of players being randomly matched to play a symmetric Cournot oligopoly game. For later convenience we start by briefly considering the traditional Cournot duopoly analysis. For a more general review on existence and stability of Cournot-Nash and Bertrand-Nash equilibria we refer to Chapter 6.

5.2.1 Traditional Cournot Duopoly Analysis

Let x_i denote the quantity supplied by player $i = 1, 2$. In addition, let $P(x_i + x_j)$, $i \neq j$, denote the inverse demand function and let $C(x_i)$ denote the cost function, which is identical for both players. We assume that $P : \mathbf{R}_+ \to \mathbf{R}_+$ is continuous, there exists $\bar{x} < \infty$ such that $P(\bar{x}) = 0$, and P is both C^2, i.e. twice continuously differentiable, and strictly decreasing on $[0, \bar{x})$. Furthermore, we assume that for all $x \in [0, \bar{x})$ it holds that $P'(x) + x_i P''(x) \leq 0$. This means that each firm's marginal revenue is decreasing in the output of the other firm. Equivalently, each firm's profit function has nonpositive cross partials with respect to its own and the other firm's output. Finally, we impose that $P(0) > C'(0)$. This condition states that each firm would like to produce at least a small quantity if it were a monopoly. With respect to the cost function we assume that $C : \mathbf{R}_+ \to \mathbf{R}_+$ is C^2 and nondecreasing.

For player i the profit resulting from the above Cournot game is given by

$$\pi_i(x_i, x_j) = P(x_i + x_j)\,x_i - C(x_i).$$

We assume that the profit function is strictly concave in the firm's own output x_i. Profit maximizing behaviour, taking the quantity supplied by the opponent as given, results in the following so-called reaction function for player i

$$x_i = R_i(x_j) = \arg\max_{x_i}\left[P(x_i + x_j)\,x_i - C(x_i)\right].$$

Due to symmetry, players i and j, $i \neq j$, have the same reaction function $R(\cdot)$. Furthermore, for both players the Cournot-Nash equilibrium quantity x^* corresponds to the solution of

$$x^* = R(x^*).$$

Since profit functions are strictly concave such a solution x^* exists. The first-order condition for a maximum is

$$P(x) + x_i P'(x) - C'(x_i) = 0.$$

Taking the total differential of the first-order condition gives

$$\left(2P'(x) + x_i P''(x) - C''(x_i)\right)\partial x_i + \left(P'(x) + x_i P''(x)\right)\partial x_j = 0.$$

Rearranging terms we find that

$$R_i'(x_j) = \frac{\partial x_i}{\partial x_j} = -\frac{P'(x) + x_i P''(x)}{2P'(x) + x_i P''(x) - C''(x_i)}.$$

Again, symmetry implies that $R_i'(\cdot) = R_j'(\cdot)$. First, notice that the derivative of the reaction function is nonincreasing. The numerator is nonpositive because marginal revenue is decreasing in the other firm's output and the denominator is strictly negative because the profit function is assumed to be strictly concave. Second, the derivative of the reaction function is continuous because both the inverse demand function and the cost function are twice continuously differentiable. Tirole (1988) states that a sufficient condition for the equilibrium x^* to be unique is that the derivative of the reaction function is unequal to 1 in absolute value over the relevant range of outputs, i.e. $|R'(x)| \neq 1$ for all $x \in [0, \bar{x})$. Since the derivative of the reaction function is continuous this implies that $|R'(x)| < 1$ for all $x \in [0, \bar{x})$ or $|R'(x)| > 1$ for all $x \in [0, \bar{x})$.

The traditional Cournot analysis refers to a static environment. However, in a dynamic setting the reaction function introduced above can be used to establish the so-called best-reply dynamics

$$x_{i,t} = R\left(x_{j,t-1}\right), i \neq j,$$

where $x_{i,t}$, $i = 1, 2$, denotes the quantity supplied by player i in period t. As shown by Rand (1978) these best-reply dynamics may converge to the Cournot-Nash equilibrium, result in an exploding path of quantities, or even result in cyclic or chaotic behaviour. In fact, the Cournot-Nash equilibrium is stable (unstable) under the best-reply dynamics if $|R'(x^*)| < 1$ ($|R'(x^*)| > 1$). The stability properties of the Cournot model under standard adjustment systems are analyzed more precisely in Hahn (1962) and Seade (1980) (see Chapter 6 for a brief review).

In part of this chapter we focus on the following linear-quadratic specification of the Cournot duopoly game. The inverse demand and cost function are given by

$$P\left(x_1 + x_2\right) = a - b\left(x_1 + x_2\right)$$

and

$$C\left(x_i\right) = cx_i - \frac{d}{2}x_i^2, \quad i = 1, 2,$$

respectively. First, in order to have a strictly concave profit function we assume that $d < 2b$. Second, we require that at all times $x_i \leq \frac{c}{d}$ since the cost function is only upward sloping if this condition holds. The fraction $\frac{c}{d}$ can be interpreted as a capacity constraint. For the above specification of the inverse demand function and cost function the reaction function is given by

$$x_i = R\left(x_j\right) = \frac{a - c - bx_j}{2b - d}.$$

It can easily be calculated that in this case the Cournot-Nash equilibrium quantity, price, and profit are equal to

$$x^* = \frac{a - c}{3b - d}, \; P\left(2x^*\right) = \frac{a\left(b - d\right) + 2bc}{3b - d}$$
$$\text{and } \pi\left(x^*, x^*\right) = \left(b - \frac{d}{2}\right)\left(\frac{a - c}{3b - d}\right)^2,$$

respectively. Furthermore, we find that

$$\frac{\partial R(q)}{\partial q} = -\frac{b}{2b-d}.$$

This implies that the Cournot-Nash equilibrium in the linear-quadratic specification of the Cournot duopoly game is stable (unstable) under the best-reply dynamics if $d < b$ ($d > b$).

5.2.2 Quantity Dynamics

As mentioned before the players act according to behavioural rules that exactly prescribe the quantity to be supplied in the Cournot game. In fact, the players can choose between K different behavioural rules $H_1, H_2, \ldots, H_K$. With each behavioural rule H_i, $i = 1, \ldots, K$, we associate an information set I_i and costs $T(I_i)$. An example of such an information set is the case where agents only need to know average industry output in the previous period. Another, more complete, information set may contain additional information about the structure of the underlying model or information about the best-reply mapping. It seems obvious that obtaining the latter information set requires more resources than the former. Therefore the costs $T(I_i)$ associated with obtaining this information set are higher.

Let $n_{i,t}$, $i = 1, \ldots, K$, denote the fraction of players using rule H_i in period t and let $x_{i,t}$, $i = 1, \ldots, K$, denote the quantity supplied by a player using rule H_i in period t. Given a certain information set I_i the corresponding behavioural rule is given by

$$x_{i,t+1} = H_i(\mathbf{X}_t, \mathbf{N}_t), \tag{5.1}$$

where

$$\mathbf{X}_t = \begin{bmatrix} x_{1,t} & x_{1,t-1} & \cdots & x_{1,t-L} \\ x_{2,t} & x_{2,t-1} & & \vdots \\ \vdots & & \ddots & \\ x_{K,t} & \cdots & & x_{K,t-L} \end{bmatrix}$$

and

$$\mathbf{N}_t = \begin{bmatrix} n_{1,t} & n_{1,t-1} & \cdots & n_{1,t-L} \\ n_{2,t} & n_{2,t-1} & & \vdots \\ \vdots & & \ddots & \\ n_{K,t} & \cdots & & n_{K,t-L} \end{bmatrix}.$$

The above specification of $\mathbf{X}_t$ and $\mathbf{N}_t$ indicates that behavioural rules have a limited memory of $L+1$ periods.[1] Note that, as the columns in $\mathbf{N}_t$ sum up to one, the last row of this matrix could be ignored. We include it, however, only for notational convenience. Together, the K heterogeneous behavioural rules constitute the quantity dynamics of the model. To make sure that the quantity dynamics are well-behaved in terms of dynamic implications we impose the following regularity conditions. Let $M_{m\times n}(S)$ denote the set of all $m \times n$ matrices with entries from the set S.

Assumption 5.1 The functions $\{H_i\}$ are all C^1, i.e. continuously differentiable. Furthermore, each behavioural rule $H_i : M_{K\times(L+1)}(\mathbf{R}_+)\times M_{K\times(L+1)}([0,1]) \to \mathbf{R}_+$, $i = 1, \ldots, K$, has the Cournot-Nash quantity x^* as its unique equilibrium quantity, i.e. $x^* = H_i(\mathbf{X}^*, \mathbf{N}_t)$, for all $\mathbf{N}_t \in M_{K\times(L+1)}([0,1])$, where $\mathbf{X}^* \in M_{K\times(L+1)}(\{x^*\})$.

The latter assumption is a weak consistency assumption implying that the Cournot-Nash equilibrium x^* (or equivalently $\mathbf{X}^*$) is the unique equilibrium quantity of the complete dynamics of the Cournot duopoly model with heterogeneous behavioural rules: given that all players have been playing the Cournot-Nash equilibrium quantity for some time, each behavioural rule specifies to play the Cournot-Nash equilibrium in the current period. Particularly, it states that x^* is the unique equilibrium of each homogeneous behavioural rule. Assumption 5.1 excludes the possibility that players "believe" in different equilibrium quantities. Allowing for this bias would introduce an additional source of instability in the model, not considered in this chapter.

Remark 5.1 The Cournot duopoly game with heterogeneous rules described above can be interpreted as a symmetric $K \times K$ matrix game with a varying payoff matrix. This is caused by the fact that the quantity that a player using a certain behavioural rule supplies, and consequently also the profit this player receives, typically varies over time.

Remark 5.2 The quantity dynamics of the model can alternatively be derived from the players having heterogeneous expectations with respect to the quantities supplied by their potential opponents in the next period.

[1] Notice that a priori it is possible that different rules have a different memory length of, say $L_i + 1$ periods, but that this can be expressed in the present notation by taking $L = \max_i L_i$.

For economic models with heterogeneous beliefs we refer to Brock and Hommes (1997) and Cabrales and Hoshi (1996).

5.2.3 Population Dynamics

We will now describe how the fractions $n_{i,t}$ evolve over time. The choice of a behavioural rule is based upon its past performance. Let $\pi_{i,j,t}(x_{i,t}, x_{j,t})$ denote the realized profit of a player supplying $x_{i,t}$ when he is matched with a player supplying $x_{j,t}$ in period t. More specifically,

$$\pi_{i,j,t}(x_{i,t}, x_{j,t}) = P(x_{i,t} + x_{j,t})\, x_{i,t} - C(x_{i,t}).$$

To keep the dynamics of the model analytically tractable, we only use the realized profit in the last period to determine the fitness measure of a behavioural rule. The probability of meeting a player supplying $x_{j,t}$ in period t is equal to $n_{j,t}$. Hence, the net average profit of behavioural rule H_i in period t is

$$U_{i,t} = \sum_{j=1}^{K} n_{j,t}\pi_{i,j,t}(x_{i,t}, x_{j,t}) - T(I_i) = \Pi^i(\mathbf{x}_t, \mathbf{n}_t) - T(I_i),$$

where $\Pi^i(\mathbf{x}_t, \mathbf{n}_t)$ is the average profit of agents using behavioural rule H_i in period t and $T(I_i)$ are the associated information costs. In the above expression $\mathbf{x}_t = (x_{1,t}, \dots, x_{K,t})'$ are the quantities corresponding to the different behavioural rules in period t and $\mathbf{n}_t = (n_{1,t}, \dots, n_{K,t})'$ are the corresponding population fractions. We assume that the fitness measures $U_{i,t}$ are publicly observable.

The fitness measures $U_{i,t}$ are used to determine the population fractions $n_{i,t+1}$. In general, such an evolutionary dynamics in discrete time, describing how the population fractions evolve, is given by

$$n_{i,t+1} = h(U_{i,t}, n_{i,t}, \mathbf{U}_{-i,t}, \mathbf{n}_{-i,t}),$$

with $\mathbf{U}_{-i,t} = (U_{1,t}, \dots, U_{i-1,t}, U_{i+1,t}, \dots, U_{K,t})'$ and $\mathbf{n}'_{-i,t} = (n_{1,t}, \dots, n_{i-1,t}, n_{i+1,t}, \dots, n_{K,t})$. To make sure that the population dynamics is well-behaved in terms of dynamic implications we assume that $h(\cdot,\cdot,\cdot,\cdot)$ is continuous, nondecreasing in the first argument, and such that the population state remains in the $(K-1)$-dimensional unit simplex Δ^K, which is a subset of the K-dimensional euclidean space $\mathbf{R}^K$. Given the fact that the dynamics remains in the unit simplex, the first regularity requirement is a sufficient condition for the existence of an equilibrium of the population dynamics. Note that dynamics in discrete time, contrary to dynamics in continuous time, automatically induce existence and uniqueness of a solution to the associated dynamical system through any initial population state in the simplex. The second regularity requirement reflects the assumption that successful behaviour reproduces more rapidly. In Section 5.3 and Section 5.4 we introduce some commonly used specific population dynamics that satisfy these conditions and figure prominently in economic applications.

The preceding discussion shows that the evolutionary game theoretic model with K behavioural rules can be transformed into a $(2K-1)$ - dimensional dynamical system. We need K equations to keep track of the quantity dynamics, i.e. the quantities supplied in the Cournot duopoly game by the players using the different behavioural rules. The other $K-1$ equations describe the population dynamics, i.e. the fractions of the population using the different behavioural rules. Consequently, the general dynamical system considered in this chapter is the following

$$\begin{aligned} x_{i,t+1} &= H_i\left(\mathbf{X}_t, \mathbf{N}_t\right) \text{ for all } i = 1, \ldots, K \\ n_{i,t+1} &= h\left(U_{i,t}, n_{i,t}, \mathbf{U}_{-i,t}, \mathbf{n}_{-i,t}\right) \text{ for all } i = 1, \ldots, K-1. \end{aligned} \tag{5.2}$$

In the examples analyzed in Section 5.3 and Section 5.6 we focus on models with $K = 2$ behavioural rules. These evolutionary models give rise to (at most) 3-dimensional dynamical systems. This is no restriction since there are relatively few dynamical phenomena that are currently understood which only occur in dimension four or more.

Remark 5.3 The definition of the fitness measure can be generalized in a straightforward manner to include the performance of the behavioural rule over the last M periods. In case there is a delay in evaluating the behavioural rules, a fitness measure $U_{i,t}$ becomes

$$\begin{aligned} U_{i,t} &= \sum_{m=0}^{M} w_{i,m} \sum_{j=1}^{K} n_{j,t-m} \pi_{i,j,t-m} - T\left(I_i\right) \\ &= \sum_{m=0}^{M} w_{i,m} \Pi^i \left(\mathbf{x}_{t-m}, \mathbf{n}_{t-m}\right) - T\left(I_i\right), \ \forall i, \end{aligned}$$

where

$$\sum_{k=0}^{M} w_{i,k} = 1, \forall i, \text{ and } w_{i,k} \geq 0, \forall i, k.$$

Numerical simulations indicate that incorporating a delay yields similar results as working with the more basic definition of a fitness measure used in this chapter.

Discrete Choice Model

This section contains a brief discussion of the discrete choice model. This model is treated extensively in Anderson, De Palma, and Thisse (1992) and Manski and McFadden (1981). Brock and Hommes (1997) also focus on the discrete choice model. Other interesting applications of the discrete choice framework can be found in Goeree (1997). Furthermore, see Chen, Friedman, and Thisse (1997) for an interesting discussion on the probabilistic choice approach.

Suppose that the utility associated with using behavioural rule H_i takes the form

$$\widetilde{U}_i = U_i + \frac{1}{\beta}\varepsilon_i,$$

where the ε_i's are IID. In fact, U_i is the deterministic part and $\frac{1}{\beta}\varepsilon_i$ captures the stochastic part of the utility of behavioural rule H_i. As explained by Anderson, De Palma, and Thisse (1992) the fraction of the population choosing behavioural rule H_i corresponds to the probability $\Pr\left(\widetilde{U}_i = \max_j \widetilde{U}_j\right)$. This captures the idea of bounded rationality since individuals do not necessarily select the rule that yields the highest utility. If the ε_i's are distributed according to the extreme value distribution this results in the so-called multinomial logit model, corresponding to the following updating of fractions

$$n_{i,t+1} = \frac{\exp\left[\beta U_{it}\right]}{\sum_{j=1}^{K} \exp\left[\beta U_{jt}\right]}, \forall i. \tag{5.3}$$

The parameter β is the intensity of choice measuring the inertia in switching behavioural rules. Notice that if $\beta = 0$ behaviour is completely random: every behavioural rule is chosen with probability $\frac{1}{K}$. If $\beta \to \infty$ the most profitable strategy is played by the complete population. In fact, β can be interpreted as a measure of the natural selection with respect to the behavioural rules.

Note that in the case of the discrete choice model the population dynamics $h(\cdot,\cdot,\cdot,\cdot)$ even remains in the interior of the unit simplex for finite β. This implies that in each time period all behaviour rules are present in the population and no behavioural rule will ever vanish (this is the so-called *no-extinction* condition). Furthermore, no new behavioural rules emerge in this model (this is the so-called *no-creation* condition). However, as we generally view all strategies as being represented in the initial distribution the no-creation condition has little bite. The equilibrium fractions are easily seen to be

$$n_i^* = \frac{\exp\left[-\beta T\left(I_i\right)\right]}{\sum_{j=1}^{K} \exp\left[-\beta T\left(I_i\right)\right]}.$$

Replicator Dynamics

Another dynamics that figures prominently in economic applications is the discrete time replicator dynamics. In this case the fraction of agents playing behavioural rule H_i in period $t+1$ becomes

$$n_{i,t+1} = \frac{n_{i,t} U_{i,t}}{\sum_{j=1}^{K} n_{j,t} U_{j,t}},$$

In this specification $U_{i,t}$ can be interpreted as the number of offspring of an individual using behavioural rule H_i in period t. Obviously, the replicator dynamics satisfies both the no-creation and the no-extinction condition. This implies that offspring increases with the payoff of a behavioural rule. As shown by Gale, Binmore and Samuelson (1995), Binmore and Samuelson (1997), and Schlag (1996), using the replicator dynamics to represent the population dynamics can be motivated in the context of a learning, aspiration, or imitation story. Furthermore, the simplicity of the replicator dynamics lends it considerable appeal. Note that at an equilibrium of these replicator dynamics we must have $n_i^* = 0$, $n_i^* = 1$, or

$$\sum_{j \in J} n_j^* U_j^* = U_i^*, \forall i \in J,$$

where $J \subseteq \{1, \ldots, K\}$ is the set of behavioural rules that are being played with positive probability in the equilibrium. This implies that U_i^* has to be the same for all $i \in J$ and this means that information costs $T\left(I_i\right)$ have to be equal for all $i \in J$. If this is the case any set of fractions $\mathbf{n}^* = \left(n_1^*, \ldots, n_K^*\right)'$, with $\sum_{j \in J} n_j^* = 1$ and $n_i^* = 0$ for $i \notin J$ is an equilibrium. The case of equal information costs is considered in Proposition 5.1 in Section 5.2.4.

In general, when information costs $T(I_i)$ are different for all behavioural rules, we have K distinct equilibria. In each of these equilibria only one behavioural rule is played with positive probability. Only the equilibrium in which the behavioural rule with the lowest information costs $T(I_i)$ is played with positive probability can be locally stable. In any other equilibrium it is profitable to choose a behavioural rule with lower information costs, since the system will remain at the Cournot-Nash equilibrium. When in an equilibrium only the behavioural rule with the lowest information costs is played with positive probability, the (in)stability of an equilibrium of the complete system follows from the (in)stability of this behavioural rule.

In the remaining part of the chapter we mainly focus on the replicator dynamics with deterministic noise, as discussed by Young and Foster (1991). The population dynamics then looks like

$$n_{i,t+1} = (1 - K\delta) \frac{n_{i,t} U_{i,t}}{\sum_{j=1}^{K} n_{j,t} U_{j,t}} + \delta, \tag{5.4}$$

where $\delta < \frac{1}{K}$ is assumed to be small. The parameter δ is called deterministic noise and its interpretation is as follows. Each period t a fraction of $K\delta$ firms leaves the market and is replaced by new firms. In their first period these new firms choose one of the existing behavioural rules randomly. A law of large numbers argument can be used to show that each behavioural rule is chosen by a fraction of $\frac{1}{K}$ of the new firms. The presence of the noise implies that equilibria have to be interior: the equilibria in which only one behavioural rule is played are destroyed. The interior equilibria have the property that $\delta < n_i^* < 1 - (K-1)\delta$ for all $i = 1, \ldots, K$. Uniqueness of the equilibrium population fractions in case the costs of the behavioural rules are different, is shown by the following Lemma.

Lemma 5.1 *Assume $T_1 > \ldots > T_K \geq 0$. Then the equilibrium population fractions of the dynamical system with the population dynamics modelled by the replicator dynamics (5.4) are unique. Furthermore, as δ approaches 0, n_K^* approaches 1 and n_i^* approaches 0 for all $i \neq K$.*

Proof. We want to show that there exists a unique solution to

$$n_i = (1 - K\delta) \frac{n_i U_i}{\sum_{j=1}^{K} n_j U_j} + \delta \tag{5.5}$$

where $n_i > 0$ for all i, $\sum_{i=1}^{n} n_i = 1$ and $U_i = \Pi^* - T_i$ is independent of the fractions. Equation (6.29) is a continuous mapping from the unit simplex into itself, so we know a fixed point exists. Now we have to show uniqueness of this fixed point. Equation (6.29) can be written as

$$(n_i - \delta) \sum_{j=1}^{K} n_j U_j = (1 - K\delta)\, n_i U_i$$

Define $c \equiv \sum_{j=1}^{K} n_j U_j$ and $b_i \equiv (1 - K\delta)\, U_i$. Given the value of c the equilibrium population fractions are uniquely determined by

$$n_i = \frac{c\delta}{c - b_i}. \tag{5.6}$$

Notice that nonnegativity of the equilibrium fractions requires that $c = \sum_{j=1}^{K} n_j U_j > (1 - K\delta)\, U_i = b_i$. Also notice that from (6.30) it follows that the fractions n_i are decreasing in c. Now suppose there are two distinct nonnegative solutions of (6.29), namely $\mathbf{n}$ and $\mathbf{m}$. It follows from (6.30) that $\sum_{j=1}^{K} n_j U_j = \sum_{j=1}^{K} m_j U_j$ implies that $\mathbf{n} = \mathbf{m}$. Now suppose that

$$c_n = \sum_{j=1}^{K} n_j U_j > \sum_{j=1}^{K} m_j U_j = c_m.$$

Given that $c_n > b_i$ and $c_m > b_i$ for all i (and therefore that all fractions n_i and m_i are positive), this implies that

$$n_i = \frac{c_n \delta}{c_n - b_i} < \frac{c_m \delta}{c_m - b_i} = m_i.$$

This leads to a contradiction since

$$1 = \sum_{i=1}^{n} n_i < \sum_{i=1}^{n} m_i = 1.$$

Hence there is a unique positive equilibrium $\mathbf{n}$.

Furthermore, multiplying n_i by U_i, aggregating over i and using (6.30) we see that c is a solution to

$$\sum_{i=1}^{K} \frac{U_i}{c - (1 - K\delta)\, U_i} = \frac{1}{\delta}.$$

As δ approaches 0, and given that $c > (1 - K\delta) U_i = b_i$, we find that $c = \sum_{j=1}^{K} n_j U_j$ has to approach $(1 - K\delta) U_K \approx U_K = \max_{i=1,\dots,K} U_i$. This implies that n_K approaches 1 and all other fractions approach 0 as δ approaches 0. ■

In this chapter we will primarily be concerned with the case $K = 2$, where there is one behavioural rule with zero information costs and one behavioural rule with positive information costs T. Some straightforward calculations show that the unique equilibrium fraction of people playing the behavioural rule with information costs T is equal to

$$n^* = \frac{1}{2T}\left(\delta\left(2\Pi^* - T\right) + T - \sqrt{\delta^2\left(2\Pi^* - T\right)^2 + (1 - 2\delta)\,T^2}\right),$$

where Π^* is the profit, exclusive of information costs, made by all the players in the equilibrium of the dynamical system. In fact, it can be shown that $\delta < n^* < \frac{1}{2}$.

There is an important difference between the discrete choice population dynamics and the replicator dynamics. The former determines fractions purely on the basis of the fitness measure, whereas the latter also incorporates the population fractions of the previous period. This implies that the population dynamics for the discrete choice model can move more rapidly: it is possible that almost the whole population switches from one rule to another in one period (provided β is sufficiently high). For the replicator dynamics, there is some extra inertia since, even if a rule performs very badly in a certain period, a relatively high fraction of the population keeps playing it if a high fraction of the population played it in the previous period. Adjustment in the population fractions under the replicator dynamics is therefore more gradual than under the discrete choice dynamics (compare the time series for the discrete choice dynamics and the replicator dynamics in Section 5.4).

5.2.4 Local Instability Results

The Cournot duopoly model with K heterogeneous behavioural rules introduced in the previous section, can be represented by a $(2K - 1)$-dimensional dynamical system. We need the first K equations to keep track of the quantity dynamics as specified in (5.1), while the remaining $K - 1$ equations describe the population dynamics given by (5.4) or (5.3). In this section we present two instability results which show that the dynamics typically does not settle down to the equilibrium.

Let $(x^*, \mathbf{n}^{r*})$, with $\mathbf{n}^{r*} = (n_1^{r*}, \ldots, n_K^{r*})'$, and $(x^*, \mathbf{n}^{d*})$, with $\mathbf{n}^{d*} = (n_1^{d*}, \ldots, n_K^{d*})'$, denote the equilibrium of the dynamical system in case the population dynamics is modeled by the replicator dynamics (5.4) and the discrete choice model (5.3), respectively. From Assumption 5.1 and Lemma 5.1 it follows that, independent of whether the population dynamics is described by the replicator dynamics (5.4) or the discrete choice model (5.3), the dynamical system has indeed a unique equilibrium.

In addition, Assumption 5.1 implies that $\mathbf{x}^* \in M_{K\times 1}(\{x^*\})$ is the unique equilibrium quantity of the Cournot duopoly model with heterogeneous behavioural rules. Consider such a model with all population fractions being fixed at $\frac{1}{K}$, i.e., $x_{i,t+1} = H_i(\mathbf{X}_t, \bar{\mathbf{N}})$ for all $i = 1, \ldots, K$, where $\bar{\mathbf{N}} \in M_{K\times(L+1)}(\{\frac{1}{K}\})$. Thus, firms stay uniformly distributed over the set of behavioural rules. The quantity dynamics can then be described as $\mathbf{x}_{t+1} = \mathbf{H}(\mathbf{X}_t, \bar{\mathbf{N}})$. We make the following assumption.

Assumption 5.2 The equilibrium quantity $\mathbf{x}^*$ of the model with all population fractions being fixed at $\frac{1}{K}$ is hyperbolic, i.e., for all i, $i = 1, \ldots, K$, the linearization of $\mathbf{x}_{t+1} = \mathbf{H}(\mathbf{X}_t, \bar{\mathbf{N}}_i)$ at $\mathbf{x}^*$ has no eigenvalues on the unit circle and x^* is locally unstable.

The above assumption means that when all firms stay uniformly distributed over the set of behavioural rules, the unique equilibrium quantity $\mathbf{x}^*$ of the resulting quantity dynamics is unstable.

The first instability result says that when all behavioural rules involve equal costs, the unstable equilibrium of the dynamical system with all population fractions being fixed at $\frac{1}{K}$ cannot be stabilized by switching of behavioural rules. Given that all behavioural rules involve equal costs, we can assume without loss of generality that $T_i = 0$ for all i, $i = 1, ..., K$.

Proposition 5.1 *Assume 5.1, 5.2 and $T_i = 0$ for every behavioural rule H_i, $i = 1, \ldots, K$. Then the equilibrium of the dynamical system with the population dynamics modelled by either the replicator dynamics (5.4), i.e., $(x^*, \mathbf{n}^{r*})$, or the discrete choice model (5.3), i.e., $(\mathbf{x}^*, \mathbf{n}^{d*})$, is unstable.*

Proof. The following analysis holds both for the population dynamics described by (5.4), in which case $\mathbf{n}^*$ and $\mathbf{N}_{i,t}$ should be interpreted as $\mathbf{n}^{r*}$ and $\mathbf{N}_{i,t}^r$, respectively, and the population dynamics described by (5.3), in which case $\mathbf{n}^*$ and $\mathbf{N}_{i,t}$ should be interpreted as $\mathbf{n}^{d*}$ and $\mathbf{N}_{i,t}^d$, respectively. Because $T_i = 0$ for all i, $i = 1, \ldots, K$, it follows immediately that the equilibrium population fractions $n_i^{r*} = n_i^{d*} = \frac{1}{K}$ for all i, $i = 1, \ldots, K$.

For all i, $i = 1, \ldots, K$, the linearization of $x_{i,t+1} = H_i(\mathbf{X}_t, \mathbf{N}_t)$ at the equilibrium $(\mathbf{x}^*, \mathbf{n}^*)$ is given by

$$\begin{aligned} dx_{i,t+1} &= \mathbf{e}^T \left[\left.\frac{\partial H_i}{\partial \mathbf{X}_t}\right|_{(x^*,\mathbf{n}^*)} \circ d\mathbf{X}_t \right] \mathbf{e} + \mathbf{e}^T \left[\left.\frac{\partial H_i}{\partial \mathbf{N}_t}\right|_{(x^*,\mathbf{n}^*)} \circ d\mathbf{N}_t \right] \mathbf{e} \\ &= \mathbf{e}^T \left[\left.\frac{\partial H_i}{\partial \mathbf{X}_t}\right|_{(x^*,\mathbf{n}^*)} \circ d\mathbf{X}_t \right] \mathbf{e} \\ &\quad + \mathbf{e}^T \left[\left.\frac{\partial H_i}{\partial \mathbf{N}_t}\right|_{(x^*,\mathbf{n}^*)} \circ \mathbf{N}_t \right] \mathbf{e} - \frac{1}{K}\mathbf{e}^T \left[\left.\frac{\partial H_i}{\partial \mathbf{N}_t}\right|_{(x^*,\mathbf{n}^*)} \right] \mathbf{e}, \end{aligned}$$

where $dx_{i,t+1} = x_{i,t+1} - x^*$, $dn_{i,t} = n_{i,t} - n_i^*$, $\mathbf{e}^T \in M_{1\times K}(\{1\})$, $\mathbf{e} \in M_{(L+1)\times 1}(\{1\})$, and $\circ$ denotes the Hadamard or entry-wise product. From assumption 5.1 it follows that

$$\left(\left.\frac{\partial H_i}{\partial \mathbf{N}_t}\right|_{(\mathbf{x}^*,\mathbf{n}^*)} \right)_{g,l} = 0 \tag{5.7}$$

for all $i, g = 1, \ldots, K$ and $l = 1, \ldots, L+1$, which implies that $dx_{i,t+1}$ can be rewritten as

$$dx_{i,t+1} = \mathbf{e}^T \left[\left.\frac{\partial H_i}{\partial \mathbf{X}_t}\right|_{(\mathbf{x}^*,\mathbf{n}^*)} \circ \delta\mathbf{X}_t \right] \mathbf{e}$$

which is exactly the linearization of $x_{i,t+1} = H_i(\mathbf{X}_t, \bar{\mathbf{N}})$ at $\mathbf{x}^*$. Now since by (5.7) the upper right block of the Jacobian of the linearized system contains only zeros, all eigenvalues of the linearization of $\mathbf{x}_{t+1} = \mathbf{H}(\mathbf{X}_t, \overline{\mathbf{N}})$ at $\mathbf{x}^*$ are eigenvalues of the linearization of the complete system at $\mathbf{x}^*$. Hence assumption 5.2 implies that the equilibrium $(\mathbf{x}^*, \mathbf{n}^*)$ of the complete dynamical system is also unstable. ■

A second instability result arises when there are different costs associated with the behavioural rules. Assume without loss of generality that $T_1 > \ldots > T_K \geq 0$. We replace Assumption 5.2 by Assumption 5.2'.

Assumption 5.2' When the whole population of firms use the cheapest behavioural rule H_K, the equilibrium quantity x^* is hyperbolic and unstable.

In an economic setting it is natural to take the cheapest behavioural rule to be some myopic or naive rule of thumb. Such a behavioural rule typically includes a time lag, does not incorporate any knowledge about the way the market works, and lacks information on the equilibrium quantities. Furthermore, a myopic rule is likely to include persistent forecast errors with respect to the behaviour of other market participants, or not to take this behaviour into account at all. It is not difficult to imagine that markets are unstable under such a behavioural rule.

Consider the dynamical system with the population dynamics represented by the replicator dynamics (5.4). For small deterministic noise δ, at the equilibrium $(\mathbf{x}^*, \mathbf{n}^{r*})$, most firms will use the cheapest behavioural rule H_K. The second instability result says that instability is inherent in such situations when more sophisticated behavioural rules are more expensive. A similar argument holds for a large intensity of choice β, in case the population dynamics is given by the discrete choice model (5.3).

Proposition 5.2 *Assume 5.1, 5.2', and* $T_1 > \ldots > T_K \geq 0$. *In case the population dynamics is modelled by the replicator dynamics (5.4) and the deterministic noise* δ *is sufficiently small, then the equilibrium* $(\mathbf{x}^*, \mathbf{n}^{r*})$ *of the dynamical system is unstable. Furthermore, in case the population dynamics is modeled by the discrete choice model (5.3) and the intensity of choice* β *is sufficiently large, then the equilibrium* $(\mathbf{x}^*, \mathbf{n}^{d*})$ *of the dynamical system is unstable.*

Proof. Consider the dynamical system with the population dynamics modeled by the replicator dynamics (5.4). According to the replicator dynamics, $n_{i,t+1} = f(\mathbf{x}_t, \mathbf{n}_t)$ for all, $i = 1, \ldots, K$. Hence the dynamics can be represented by

$$x_{i,t+1} = H_i(\mathbf{X}_t, \mathbf{N}_t^r) = \tilde{H}_i\left(\mathbf{x}_t, \ldots, \mathbf{x}_{t-L}, \mathbf{n}_{t-L}^r\right), \tag{5.8}$$

where $\mathbf{x}_{t-l} = (x_{1,t-l}, \ldots, x_{K,t-l})'$ for all l, $l = 0, \ldots, L$, $\mathbf{n}_{t-L}^r = (n_{1,t-L}^r, \ldots, n_{K,t-L}^r)'$. First, notice that including the population fraction $n_{K,t-L}$ in (5.8) is redundant since $\sum_{i=1}^K n_{i,t-L} = 1$.Second, notice that (5.8) is of dimension $(L+2)K - 1$. By Assumption 5.1 we know that for all i, $i = 1, \ldots, K$, it holds that

$$\tilde{H}_i\left(\mathbf{x}^*, \ldots, \mathbf{x}^*, \mathbf{n}_{t-L}^r\right) = x^*$$

for any $\mathbf{n}_{t-L}^r$. Thus, the partial derivative

$$\frac{\partial \tilde{H}_i\left(\mathbf{x}^*, \ldots, \mathbf{x}^*, \mathbf{n}_{t-L}^r\right)}{\partial n_{j,t-L}} = 0$$

for all $j = 1, \ldots, K-1, i = 1, \ldots, K$. Let $\mathbf{n}^{r*}(\delta) = (n_1^{r*}(\delta), \ldots, n_K^{r*}(\delta))'$ denote the equilibrium population fractions, and consider the behavioural rules H_i, $i = 1, \ldots, K$, such that

$$x_{i,t+1} = H_i(\mathbf{X}_t, \mathbf{N}^{r*}(\delta)) = \widehat{H}_i(\mathbf{x}_t, \ldots, \mathbf{x}_{t-L}), \tag{5.9}$$

with $\mathbf{N}^{r*}(\delta) = [\mathbf{n}^{r*}(\delta), \ldots, \mathbf{n}^{r*}(\delta)]_{K\times(L+1)}$, describing the quantity dynamics when the population fractions of firms using behavioural rules $H_1, \ldots, H_K$ are fixed at the equilibrium population fractions of the dynamical system. Using (5.9) it can be shown that the Jacobian of (5.8), evaluated at the equilibrium has $K-1$ eigenvalues $\lambda_1, \ldots, \lambda_{K-1}$ such that $0 < \lambda_1, \ldots, \lambda_{K-1} < 1$, while the other $(L+1)K$ eigenvalues are equal to the eigenvalues of (5.9), evaluated at the equilibrium.

For δ small, the equilibrium population fraction using the cheapest behavioural rule H_K approaches one, i.e., $n_K^{r*}(\delta) \approx 1$. Now consider the system describing the quantity dynamics when the whole population of firms uses behavioural rule H_K. By Assumption 5.2', the equilibrium quantity x^* of this system is unstable. Since, for δ small, the system given by (5.9) gets C^1-close to the system in which the whole population of firms uses behavioural rule H_K, we conclude that for δ sufficiently small, at the equilibrium, the system specified by (5.9) is unstable. This implies that the dynamical system with the population dynamics modeled by the replicator dynamics (5.4) is unstable at the equilibrium.

A similar argument as above can be used to show that for β sufficiently large, the dynamical system with the population dynamics modeled by the discrete choice model (5.3) is also unstable at the equilibrium.

■

5.3 Best-Reply versus Rational Players

In this section we consider a simple, but typical, example of the general model introduced in Section 5.2. In fact, we analyze the case where there are $K = 2$ behavioural rules available in the population. Namely, a best-reply behavioural rule and a so-called rational behavioural rule. Let n_t be the fraction of rational players in period t and let x_t and y_t denote the quantities supplied by the best-reply players and the rational players in period t, respectively.

With respect to the best-reply players we assume that they know average industry output in the previous period. Since average industry output in period t is equal to $\overline{x}_t = n_t y_t + (1 - n_t) x_t$, the best-reply players supply the quantity

$$x_{t+1} = R(\bar{x}_t) = R(n_t y_t + (1 - n_t) x_t)$$

in period $t+1$. From the above expression it follows that the introduction of a certain amount of rational players may stabilize the best-reply dynamics. In fact, we have $\frac{dR(\bar{x}_t)}{dx_t} = (1 - n_t) R'(\overline{x}_t)$. Consequently, if the fraction n_t is large enough the resulting best-reply dynamics may become stable, even though the original best-reply dynamics is not (that is, when $|R'(x^*)| > 1$).

Rational players reason more subtle than best-reply players. We assume that at time $t+1$ they know the fraction of people playing according to the rational behavioural rule in that period. Rational players can therefore also be interpreted as perfect foresight players. In addition, they are able to calculate the quantity x_{t+1} supplied by the best-reply players in period $t+1$. Now suppose the inverse demand function is linear. The optimal strategy of the rational players then is implicitly defined as follows

$$y_{t+1} = R(n_{t+1} y_{t+1} + (1 - n_{t+1}) x_{t+1}).$$

The strategy of the rational players is something like a Nash equilibrium in a game which is "contaminated" with a number of best-reply players. Because of the extensive computational and informational requirements of the rational behavioural rule it seems reasonable to associate costs $T > 0$ with this rule. Since we assumed $R(\cdot)$ to be downward sloping the rational behavioural rule can be determined implicitly as

$$y_{t+1} = G(x_{t+1}, n_{t+1}).$$

Note that both rules discussed above are unbiased as they have the Cournot-Nash equilibrium quantity x^* as their unique equilibrium quantity. Furthermore, x^* is also the unique equilibrium quantity of the complete dynamics of the model with heterogeneous behavioural rules. In fact, this dynamical system is given by

$$\begin{aligned} x_{t+1} &= f(x_t, y_t, n_t) = R(n_t y_t + (1 - n_t) x_t) \\ y_{t+1} &= g(x_t, y_t, n_t) = G(f(x_t, y_t, n_t), h(x_t, y_t, n_t)) \\ n_{t+1} &= h(x_t, y_t, n_t), \end{aligned}$$

where the mapping $h(x_t, y_t, n_t)$ represents the population dynamics. With respect to the above dynamical system it should be noted that

$$x_{t+1} = R(n_t y_t + (1 - n_t) x_t) = y_t.$$

This means that the only difference between the rational strategy and the best-reply strategy is that the latter always lags one period behind. Therefore, we can reduce the dimension of the dynamical system by one and write everything in terms of x_t. The resulting 2-dimensional dynamical system is

$$\begin{aligned} x_{t+1} &= R\left(n_t G\left(x_t, n_t\right) + \left(1 - n_t\right) x_t\right) \\ n_{t+1} &= h\left(x_t, G\left(x_t, n_t\right), n_t\right). \end{aligned} \tag{5.10}$$

In the following two sections we will analyze this two-dimensional nonlinear dynamical system. Section 5.4 discusses local bifurcation results for the discrete choice dynamics and the replicator dynamics and in Section 5.5 we show the existence of a homoclinic bifurcation for the replicator dynamics and the linear-quadratic specification of the model.

5.4 Local Bifurcation Analysis

In this section we analyze the model introduced in the previous section, for two types of population dynamics, the discrete choice model and the replicator dynamics. The analysis in this section is local, that is, we investigate the local stability of the evolutionary model when the evolutionary pressure (as expressed by the parameter β for the discrete choice dynamics and the parameter δ for the replicator dynamics) increases. We will also show the results from some simulations, to get an intuition as to what happens globally. Before specifying the population dynamics $h\left(x_t, G\left(x_t, n_t\right), n_t\right)$ we make the following observation with respect to the dynamical system (5.10).

Lemma 5.2 *Consider the dynamical system (5.10). The eigenvalues of the Jacobian matrix of the linearized system, evaluated at the equilibrium* (x^*, n^*), *are* $\lambda_1 = \frac{(1-n^*)R'(x^*)}{1-n^* R'(x^*)}$ *and* $\lambda_2 = \left.\frac{\partial h(x_t, G(x_t, n_t), n_t)}{\partial n_t}\right|_{(x^*, n^*)}$.

Proof. To determine the eigenvalues of the Jacobian matrix evaluated in the equilibrium (x^*, n^*), we consider the entries in the first row of this matrix. It is easily seen that these entries are given by

$$J_{11} = \left(n^* G_1\left(x^*, n^*\right) + \left(1 - n^*\right)\right) R'\left(x^*\right) \text{ and } J_{12} = n^* G_2\left(x^*, n^*\right) R'\left(x^*\right),$$

where G_i is the derivative of the function $G\left(\cdot, \cdot\right)$ with respect to its i'th variable. To compute both $G_1\left(x^*, n^*\right)$ and $G_2\left(x^*, n^*\right)$ we need to take the total differential of

$$y_{t+1} = R\left(n_t y_{t+1} + (1-n_{t+1})\, x_{t+1}\right),$$

which results in

$$\begin{aligned} \partial y_{t+1} &= R'\left(n_{t+1}y_{t+1} + (1-n_{t+1})\,x_{t+1}\right) \\ &\quad \times \left((1-n_{t+1})\,\partial x_{t+1} + n_{t+1}\partial y_{t+1} + (y_{t+1}-x_{t+1})\,\partial n_{t+1}\right). \end{aligned}$$

Rearranging terms gives

$$\frac{\partial y_{t+1}}{\partial x_{t+1}} = G_1\left(x_{t+1}, n_{t+1}\right) = \frac{(1-n_{t+1})\,R'\left(n_{t+1}y_{t+1} + (1-n_{t+1})\,x_{t+1}\right)}{1-n_{t+1}R'\left(n_{t+1}y_{t+1} + (1-n_{t+1})\,x_{t+1}\right)}$$

and

$$\frac{\partial y_{t+1}}{\partial n_{t+1}} = G_2\left(x_{t+1}, n_{t+1}\right) = \frac{(y_{t+1}-x_{t+1})\,R'\left(n_{t+1}y_{t+1} + (1-n_{t+1})\,x_{t+1}\right)}{1-n_{t+1}R'\left(n_{t+1}y_{t+1} + (1-n_{t+1})\,x_{t+1}\right)}.$$

Evaluating the above expressions in the equilibrium (x^*, n^*) results in

$$G_1\left(x^*, n^*\right) = \frac{(1-n^*)\,R'(x^*)}{1-n^*R'(x^*)} \text{ and } G_2\left(x^*, n^*\right) = 0.$$

With respect to the Jacobian matrix this implies

$$J_{11} = \left(n^* \frac{(1-n^*)\,R'(x^*)}{1-n^*R'(x^*)} + (1-n^*)\right) R'(x^*) = \frac{(1-n^*)\,R'(x^*)}{1-n^*R'(x^*)},$$

and $J_{12} = 0$. Consequently, the eigenvalues of the Jacobian matrix of the linearized system, evaluated in the equilibrium (x^*, n^*), are $\lambda_1 = J_{11} = \frac{(1-n^*)R'(x^*)}{1-n^*R'(x^*)}$ and $\lambda_2 = J_{22} = \left.\frac{\partial h(x_t, G(x_t,n_t),n_t)}{\partial n_t}\right|_{(x^*,n^*)}$. ■

5.4.1 Discrete Choice Dynamics

Suppose that the population fractions of players using a certain behavioural rule are updated according to the discrete choice model. In this case the dynamical system (5.10) becomes

$$\begin{aligned} x_{t+1} &= R\left(n_t G\left(x_t, n_t\right) + (1-n_t)\, x_t\right) \qquad (5.11) \\ n_{t+1} &= h\left(x_t, G\left(x_t, n_t\right), n_t\right) \\ &= \frac{1}{1+\exp\left[\beta\left(\Pi^b\left(x_t, G\left(x_t, n_t\right), n_t\right) - \Pi^r\left(x_t, G\left(x_t, n_t\right), n_t\right) + T\right)\right]}, \end{aligned}$$

where

$$\begin{aligned}\Pi^r(x_t, G(x_t, n_t), n_t) &= n_t\pi(G(x_t, n_t), G(x_t, n_t)) \\ &\quad + (1 - n_t)\pi(G(x_t, n_t), x_t)\end{aligned}$$

and

$$\Pi^b(x_t, G(x_t, n_t), n_t) = n_t\pi(x_t, G(x_t, n_t)) + (1 - n_t)\pi(x_t, x_t)$$

are the average profit of the rational players and the best-reply players in period t, respectively. We will study what happens to the dynamics of (5.11) as the intensity of choice β increases. Note that by definition we have that $\Pi^r(x_t, G(x_t, n_t), n_t) \geq \Pi^b(x_t, G(x_t, n_t), n_t)$.

Local Stability Analysis

The first step in analyzing the behaviour of the dynamical system (5.11) is to investigate the local stability properties of the equilibrium. These properties are summarized in Proposition 5.3.

Proposition 5.3 *Consider the dynamical system (5.11). Let $R'(x^*) < -1$. The equilibrium of (5.11) is (x^*, n^*), where x^* is the Cournot-Nash equilibrium quantity and $n^* = \frac{1}{1+\exp[\beta T]}$. Then given T there exists a value β^* of β such that the equilibrium (x^*, n^*) is locally stable for all $\beta < \beta^*$ and unstable for all $\beta > \beta^*$. Moreover, $\beta^* = \frac{1}{T}\ln\left[\frac{R'(x^*)-1}{R'(x^*)+1}\right]$.*

Proof. Straightforward calculations show that (x^*, n^*) is the equilibrium of (5.11). According to Lemma 5.2 the eigenvalues of (5.11) are $\lambda_1 = \frac{(1-n^*)R'(x^*)}{1-n^*R'(x^*)}$ and $\lambda_2 = \left.\frac{\partial h(x_t, G(x_t, n_t), n_t)}{\partial n_t}\right|_{(x^*, n^*)}$. We start by considering the eigenvalue λ_2. Note that $h(x_t, G(x_t, n_t), n_t)$ only depends upon n_t through the profit functions. Furthermore, it holds that

$$\begin{aligned}\frac{\partial \Pi^r(x_t, G(x_t, n_t), n_t)}{\partial n_t} &= n_t\left(\frac{\partial \pi(y, y)}{\partial y}G_2 + \frac{\partial \pi(y, y)}{\partial x}G_2\right) \\ &\quad + (1 - n_t)\frac{\partial \pi(y, x)}{\partial y}G_2 \\ &\quad + \pi(G(x_t, n_t), G(x_t, n_t)) - \pi(G(x_t, n_t), x_t)\end{aligned}$$

which is equal to 0 in equilibrium because $G_2(x^*, n^*) = 0$. A similar argument can be used to show that

$$\left.\frac{\partial \Pi^{br}(x_t, G(x_t, n_t), n_t)}{\partial n_t}\right|_{(x^*, n^*)} = 0.$$

From the above reasoning it follows immediately that $\lambda_2 = 0$. As a result, we have that the eigenvalue λ_1 is the only nonzero eigenvalue. Hence, the equilibrium is locally stable (unstable) if

$$\lambda_1 = \frac{(1-n^*) R'(x^*)}{1-n^* R'(x^*)} > (<) -1.$$

Since $n^* = \frac{1}{1+\exp[\beta T]}$ this condition is equivalent to $0 < \beta < \beta^*$ $(\beta > \beta^*)$, where $\beta^* = \frac{1}{T} \ln \left[\frac{R'(x^*)-1}{R'(x^*)+1} \right]$. ∎

Proposition 5.3 states that the equilibrium of (5.11) loses stability for high values of the intensity of choice β. A high intensity of choice means that there is little inertia in switching between behavioural rules. As a result, most players will switch to the myopic behavioural rule whenever the dynamical system is close to the equilibrium. Namely, close to the equilibrium the best-reply rule results in a higher profit because there are no information costs associated with this rule. Since the equilibrium is unstable under this myopic behavioural rule we are left with the above result.

Note that the equilibrium of (5.11) is locally stable for all values of the intensity of choice $\beta > 0$ in case $-1 \leq R'(x^*) \leq 0$. Furthermore, in case information costs $T = 0$ we have $\lambda_1 = \frac{\frac{1}{2}R'(x^*)}{1-\frac{1}{2}R'(x^*)}$. This means that in the model without information costs the equilibrium is stable for all values $\beta > 0$.

Finally, Proposition 5.3 could be reformulated with the roles of the intensity of choice β and the information costs T interchanged. In fact, given β there exists a value $T^* = \frac{1}{\beta} \ln \left[\frac{R'(x^*)-1}{R'(x^*)+1} \right]$ of T such that the equilibrium is locally stable for all $T < T^*$ and unstable for all $T > T^*$.

A Bifurcation Scenario

This section deals with the linear-quadratic specification of the Cournot duopoly model. For later convenience we rewrite the dynamical system in terms of deviations $X_t = x_t - x^*$ and $Y_t = y_t - x^*$ from the Cournot-Nash equilibrium quantity x^*. It can easily be shown that

$$Y_t = G(X_t, n_t) = -\frac{b(1-n_t)}{(2+n_t)b-d} X_t.$$

Consequently, the linear-quadratic specification of the dynamical system (5.11) written in deviations becomes

$$\begin{aligned} X_{t+1} &= -\frac{b\,(1-n_t)}{(2+n_t)\,b-d}X_t \\ n_{t+1} &= \frac{1}{1+\exp\left[\beta\left(T-\left(b-\frac{d}{2}\right)\left(\frac{3b-d}{(2+n_t)b-d}\right)^2 X_t^2\right)\right]}. \end{aligned} \tag{5.12}$$

Write F for the 2-dimensional map (5.12). The derivation of the profit functions for the linear-quadratic specification of the model, which are used to determine the population dynamics of (5.12), can be found in Appendix 5.8.1. We have to keep in mind that production levels are always nonnegative and smaller than the capacity constraint, that is $-x^* \leq X_t \leq \frac{c}{d} - x^*$ or, equivalently, $-\frac{a-c}{3b-d} \leq X_t \leq \frac{c}{d} - \frac{a-c}{3b-d}$. An important feature of the dynamical system (5.12) is that it is symmetric with respect to the line $X_t = 0$. This property will prove to be convenient later on. In fact, it enables us to determine the secondary bifurcation analytically.

The model with best-reply versus rational players, discrete choice dynamics, and the linear-quadratic specification of the Cournot duopoly game is, upon a transformation of variables, mathematically equivalent to the model studied by Brock and Hommes (1997). They study the evolution of equilibrium prices in a cobweb model with rational versus naive expectations. The state variable in their model is the price p_t and the parameters (next to the intensity of choice β and information costs T) are marginal supply $\widehat{b}$ and marginal demand B. In particular, if we transform the state variable p_t according to $X_t = \sqrt{\frac{2}{2b-d}}p_t$ and additionally assume that $\widehat{b} = b$ and $B = 2b - d$ we end up exactly with the model presented here. The equivalence is shown in more detail in Appendix 5.8.2.

We will now study how the dynamical behaviour of (5.12) changes as β increases. Proposition 5.4 states what happens when the equilibrium (X^*, n^*), with $X^* = 0$ and $n^* = \frac{1}{1+\exp[\beta T]}$, loses stability.

Proposition 5.4 *Consider the dynamical system (5.12). Let $b < d < 2b$. Then the primary bifurcation, i.e. a period doubling bifurcation, occurs at $\beta^* = \frac{1}{T}\ln\left[\frac{3b-d}{d-b}\right]$. In fact, for values $\beta < \beta^*$ the equilibrium $\left(0, \frac{1}{1+\exp[\beta T]}\right)$ is locally stable and for values $\beta > \beta^*$ the equilibrium is* *unstable and a locally stable period 2 cycle exists. This period 2 cycle is given by*

$$\left\{ \left(\sqrt{\frac{(\beta-\beta^*)T}{2(2b-d)\beta}}, \frac{d-b}{2b} \right)', \left(-\sqrt{\frac{(\beta-\beta^*)T}{2(2b-d)\beta}}, \frac{d-b}{2b} \right)' \right\}.$$

The secondary bifurcation, where the period 2 cycle loses stability, occurs at $\beta^{**} = \beta^* + \frac{b}{(d-b)T}$.

Proof. First, the local stability result and the primary bifurcation value follow directly from Proposition 5.3. At $\beta = \beta^*$ it holds that the eigenvalues of the Jacobian matrix, evaluated at the equilibrium, are equal to $\lambda_1 = -1$ and $\lambda_2 = 0$. Consequently, a period doubling bifurcation occurs. Second, due to the aforementioned symmetry we are looking for a period 2 cycle of the form

$$\{(X_1, n)', (X_2, n)'\},$$

where

$$X_2 = \frac{-b(1-n)}{(2+n)b-d} X_1.$$

Since the same equality must hold with X_1 and X_2 interchanged, we get

$$\frac{-b(1-n)}{(2+n)b-d} = -1,$$

implying $n = \frac{d-b}{2b}$. Writing $X = X_1 = -X_2 > 0$ and using the second equation of the dynamical system (5.12) we find that X is the positive solution of

$$n = \frac{1}{1+\exp\left[\beta\left(T + 2b(2n-1)X^2\right)\right]}.$$

Substituting $n = \frac{d-b}{2b}$ and solving this equation gives the solution provided in Proposition 5.4. Let F^2 denote the second iterate of the map F, i.e. the map F composed with itself two times. Third, to study the stability of this period 2 cycle we consider the Jacobian matrix of F^2, evaluated in the period 2 cycle $\{(X_1, n)', (X_2, n)'\}$. This Jacobian matrix $\mathbf{J}_{F^2}$ is given by

$$\begin{pmatrix} 1 - \frac{2(d-b)T(\beta-\beta^*)}{b} & \sqrt{\frac{8(\beta-\beta^*)T}{(2b-d)\beta}} \frac{b-(d-b)(\beta-\beta^*)T}{3b-d} \\ -\xi\sqrt{\frac{\beta(\beta-\beta^*)T}{2(2b-d)}} & -\frac{T(\beta-\beta^*)(d-b)(2b-(d-b)(\beta-\beta^*)T)}{b^2} \end{pmatrix},$$

where $\xi = \frac{(d-b)(2b-d)(3b-d)(b-(d-b)(\beta-\beta^*)T)}{b^3}$. It can be checked that the eigenvalues of this matrix are complex conjugates and that the determinant of $\mathbf{J}_{F^2}$ is given by

$$\det \mathbf{J}_{F^2} = \frac{(d-b)^2(\beta-\beta^*)^2 T^2}{b^2}.$$

At β^{**} the above determinant is equal to 1 and both eigenvalues lie on the unit circle. In fact, both eigenvalues are equal to -1. Finally, because $\det \mathbf{J}_{F^2}$ is increasing in β the period 2 cycle is locally stable if $\beta < \beta^{**}$ and unstable if $\beta > \beta^{**}$. ■

Proposition 5.4 indicates a source of complicated dynamics for intermediate values of the intensity of choice β. Namely, at β^{**} a so-called $1:2$ strong resonance Hopf bifurcation occurs. An important feature of this Hopf bifurcation is the emergence of four period 4 cycles, two locally stable and two unstable. Therefore, we have coexistence of low periodic attractors. It depends upon the initial state (X_0, n_0) to which of the two stable period 4 cycles, which are symmetric with respect to the line $X_t = 0$, the equilibrium path settles down. The basin of attraction of an attractor is the set of initial states (X_0, n_0) converging to the attractor. In case the dynamical system exhibits small stochastic noise, coexisting attractors can be a source of complicated dynamical behaviour since orbits may jump irregularly from one basin of attraction to the other. For a more detailed discussion of this special Hopf bifurcation we refer to Brock and Hommes (1997).

For large values of the intensity of choice β, the dynamical behaviour can be complicated due to the occurrence of strange attractors. Figure 5.1 shows the attractors of the model for the parameter values $a = 17$, $b = 1$, $c = 10$, $d = \frac{5}{4}$, $T = 1$, and for different values of β. For this numerical example the primary period doubling bifurcation occurs at $\beta^* = \ln 7$ and the $1:2$ strong resonance Hopf bifurcation of the period 2 cycle occurs at $\beta^{**} = 4 + \ln 7$. As β increases further, apparently both stable period 4 cycles turn into 4-piece chaotic attractors after a cascade of infinitely many period doubling bifurcations. One of these 4-piece chaotic attractors is depicted in Figure 5.1. Like the two stable period 4 cycles, the two coexisting 4-piece chaotic attractors are also symmetric with respect to the line $X_t = 0$. As β increases even further, our numerical simulations show the existence of a strange attractor.

Figure 5.2 shows the time series of X_t and n_t for the dynamical system associated with the numerical example when β equals 25. These time

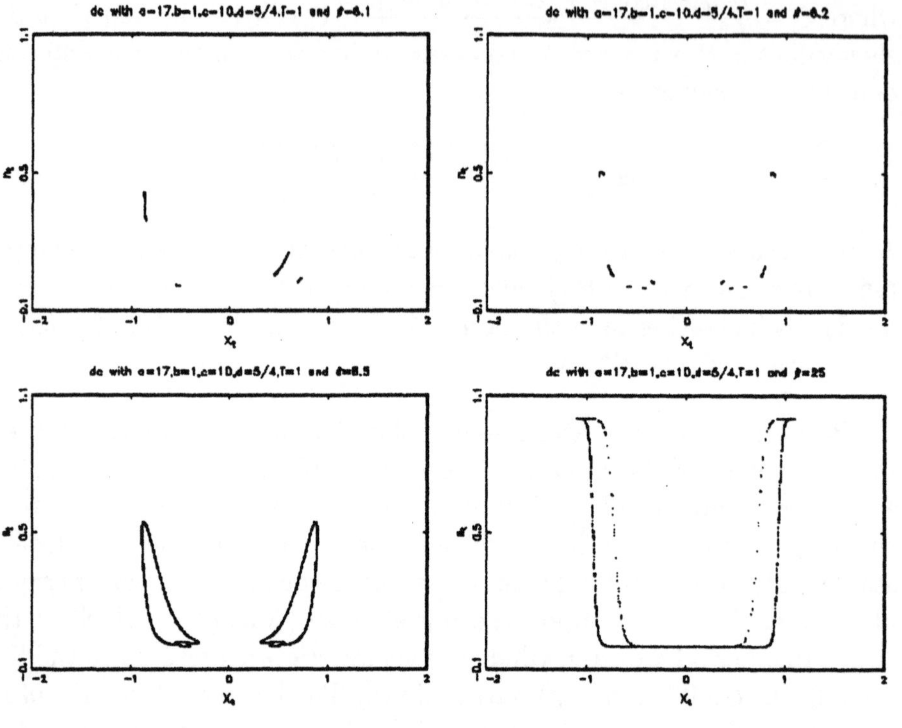

Figure 5.1: Attractors for the model with best-reply versus rational players and discrete choice dynamics with $a = 17$, $b = 1$, $c = 10$, $d = 1.25$, $T = 1$ and a) $\beta = 8.1$, b) $\beta = 8.2$, c) $\beta = 8.5$ and d) $\beta = 25$.

series clearly reveal what is going on. For some time almost all agents use the best-reply strategy, i.e. n_t is close to 0, and the production levels are close to the Cournot-Nash equilibrium, i.e. X_t is close to 0. However, since the best-reply dynamics are unstable, the dynamics move away from the equilibrium and output begins to fluctuate erratically. At a certain time they move so wildly that it becomes worthwhile to spend money on the rational behavioural rule. Then almost all agents suddenly switch behavioural rules and n_t becomes close to 1. This stabilizes the system and output returns to its equilibrium level, which in turn makes the best-reply strategy more profitable. As a result, n_t converges to 0 very quickly again.

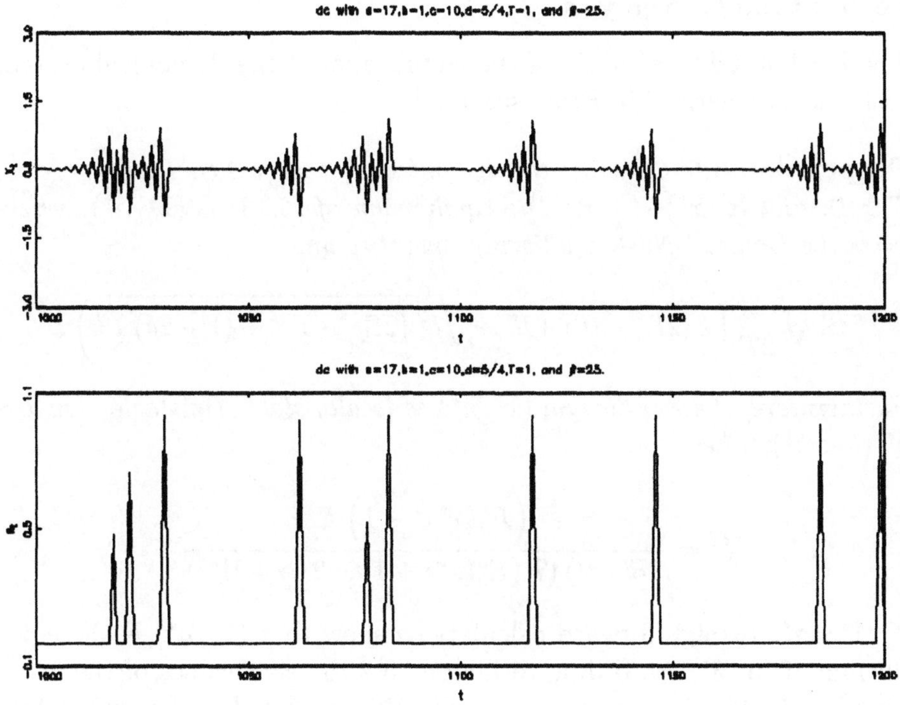

Figure 5.2: Time series of quantities and fraction of rarional players for the model with best-reply versus rational players and discrete choice dynamics with $a = 17$, $b = 1$, $c = 10$, $d = 1.25$, $T = 1$ and $\beta = 25$.

5.4.2 Replicator Dynamics

Now we consider the case where the population dynamics is modelled according to the replicator dynamics with deterministic noise instead of the discrete choice model. This transforms the dynamical system (5.10) into

$$\begin{aligned} x_{t+1} &= R\left(n_t G\left(x_t, n_t\right) + \left(1 - n_t\right) x_t\right) \qquad (5.13)\\ n_{t+1} &= \frac{n_t\left(1 - 2\delta\right)\left(\Pi^r\left(x_t, G\left(x_t, n_t\right), n_t\right) - T\right)}{n_t\left(\Pi^r\left(x_t, G\left(x_t, n_t\right), n_t\right) - T\right) + \left(1 - n_t\right)\Pi^b\left(x_t, G\left(x_t, n_t\right), n_t\right)} \\ &\quad + \delta. \end{aligned}$$

In this section we study what happens to the dynamical behaviour of (5.13) as the deterministic noise δ goes to zero.

Local Stability Analysis

The local stability analysis of the equilibrium of the dynamical system (5.13) is summarized in Proposition 5.5.

Proposition 5.5 *Consider the dynamical system (5.13). Let* $0 < \delta < \frac{1}{2}$, $T > 0$, *and* $R'(x^*) < -1$. *The equilibrium of (5.13) is* (x^*, n^*), *where* x^* *is the Cournot-Nash equilibrium quantity and*

$$n^* = \frac{1}{2T}\left(\delta\left(2\Pi^* - T\right) + T - \sqrt{\delta^2\left(2\Pi^* - T\right)^2 + \left(1 - 2\delta\right)T^2}\right).$$

Furthermore, the equilibrium (x^*, n^*) *is locally stable (unstable) for* $\delta > \delta^*$ $(\delta < \delta^*)$, *where*

$$\delta^* = \frac{\left(R'(x^*)^2 - 1\right)T}{2R'(x^*)\left(T\left(R'(x^*) + 1\right) - 2\left(\alpha + \Pi^*\right)\right)}.$$

Proof. Straightforward calculations show that (x^*, n^*) is the equilibrium of (5.13). According to Lemma 5.2 the eigenvalues of the Jacobian matrix of the dynamical system (5.10), evaluated at the equilibrium (x^*, n^*), are $\lambda_1 = \frac{(1-n^*)R'(x^*)}{1-n^*R'(x^*)}$ and $\lambda_2 = \left.\frac{\partial h(x_t, G(x_t, n_t), n_t)}{\partial n_t}\right|_{(x^*, n^*)}$. First, consider the eigenvalue λ_2. In case the population dynamics are represented by the replicator dynamics with deterministic noise it holds that

$$0 < \lambda_2 = (1 - 2\delta)\frac{(\Pi^*)(\Pi^* - T)}{(\Pi^* - n^*T)^2} < \frac{(\Pi^*)(\Pi^* - T)}{(\Pi^* - n^*T)^2} < \frac{(\Pi^*)(\Pi^* - T)}{\left(\Pi^* - \frac{1}{2}T\right)^2} < 1.$$

Note that in the above derivation we use $\delta < n^* < \frac{1}{2}$. Second, because the eigenvalue λ_2 is always strictly between 0 and 1, the equilibrium (x^*, n^*) is locally stable (unstable) if

$$\lambda_1 = \frac{(1 - n^*)R'(x^*)}{1 - n^*R'(x^*)} > (<) - 1.$$

Substituting the expression for n^* and solving for the deterministic noise δ gives the condition as stated in Proposition 5.5. ■

Proposition 5.5 says that the equilibrium of (5.13) loses stability for low values of the deterministic noise δ. A low deterministic noise means that there are hardly any players who choose a behavioural rule at random. As a result, almost all players will use the best-reply behavioural rule whenever the dynamical system is close to the equilibrium. Namely, close to the equilibrium the best-reply rule results in a higher profit because there are no information costs associated with this rule. Since the equilibrium is unstable under this best-reply behavioural rule we are left with the result mentioned in Proposition 5.5.

Note that the equilibrium (x^*, n^*) of (5.13) is locally stable for all values of the deterministic noise δ if $-1 \leq R'(x^*) \leq 0$. Furthermore, in case there are no information costs the equilibrium is stable for all values of the deterministic noise $\delta > 0$.

Again, Proposition 5.5 can be reformulated with the roles of the deterministic noise δ and the information costs T interchanged. In fact, given δ there exists a value

$$T^* = -\frac{4\delta R'(x^*)\,\Pi^*}{R'(x^*)\left(R'(x^*) - 2\delta\left(R'(x^*) + 1\right)\right) - 1}$$

of T such that the equilibrium is locally stable for all $T < T^*$ and unstable for all $T > T^*$.

A Bifurcation Scenario

In this section we consider the linear-quadratic specification of the Cournot duopoly model where the replicator dynamics with deterministic noise describes the population dynamics. We have again written the model in terms of deviations from the Cournot-Nash quantity. Straightforward calculations show that the model then becomes

$$\begin{aligned} X_{t+1} &= -\frac{b(1-n_t)}{(2+n_t)b-d}X_t \\ n_{t+1} &= (1-2\delta)\frac{n_t\left(\Pi^r(X_t,n_t)-T\right)}{n_t\left(\Pi^r(X_t,n_t)-T\right)+(1-n_t)\Pi^b(X_t,n_t)}+\delta, \end{aligned} \tag{5.14}$$

with

$$\begin{aligned} \Pi^b(X_t,n_t) &= \left(b-\frac{1}{2}d\right)\left(\frac{a-c}{3b-d}\right)^2 \\ &\quad -b(1-n_t)\left(\frac{a-c}{3b-d}\right)\left(\frac{2b-d}{(2+n_t)b-d}\right)X_t \\ &\quad -\left(b-\frac{1}{2}d\right)\left(\frac{(4-n_t)b-d}{(2+n_t)b-d}\right)X_t^2 \end{aligned}$$

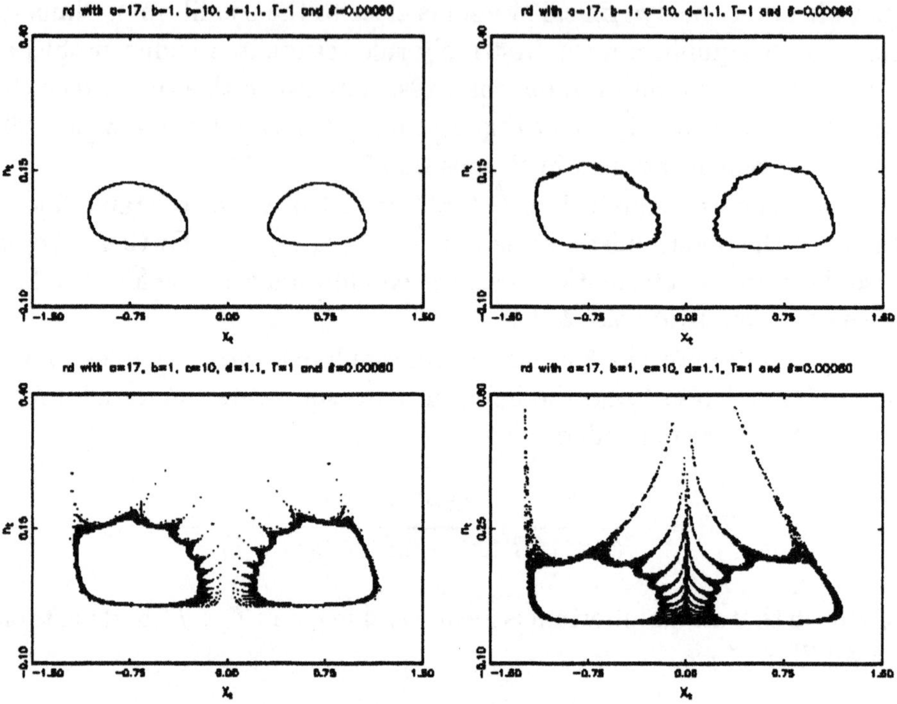

Figure 5.3: Attractors for the model with best-reply versus rational players and replicator dynamics with $a = 17$, $b = 1$, $c = 10$, $d = 1.1$, $T = 1$ and a) $\delta = 0.00080$, b) $\delta = 0.00066$, c) $\delta = 0.00064$ and d) $\delta = 0.00060$.

and

$$\Pi^r (X_t, n_t) = \left(b - \frac{1}{2}d\right) \left(\frac{b(1-n_t)}{(2+n_t)\,b-d}X_t - \frac{a-c}{3b-d}\right)^2 .$$

We shall refer to this dynamical system as $(X_{t+1}, n_{t+1}) = F(X_t, n_t)$. Note that the quantity dynamics of the current model are represented by the same equation as the quantity dynamics in (5.12). Contrary to the dynamical system (5.12), where the discrete choice model describes the population dynamics, the current linear-quadratic specification does not exhibit any symmetry. This is due to the fact that for the discrete choice dynamics the population dynamics only depends upon the *difference* in profits (which is a quadratic function in X) whereas for the replicator dynamics the *level* of profits also matters. Generically the secondary bifurcation is a Hopf bifurcation, resulting in an invariant set consisting

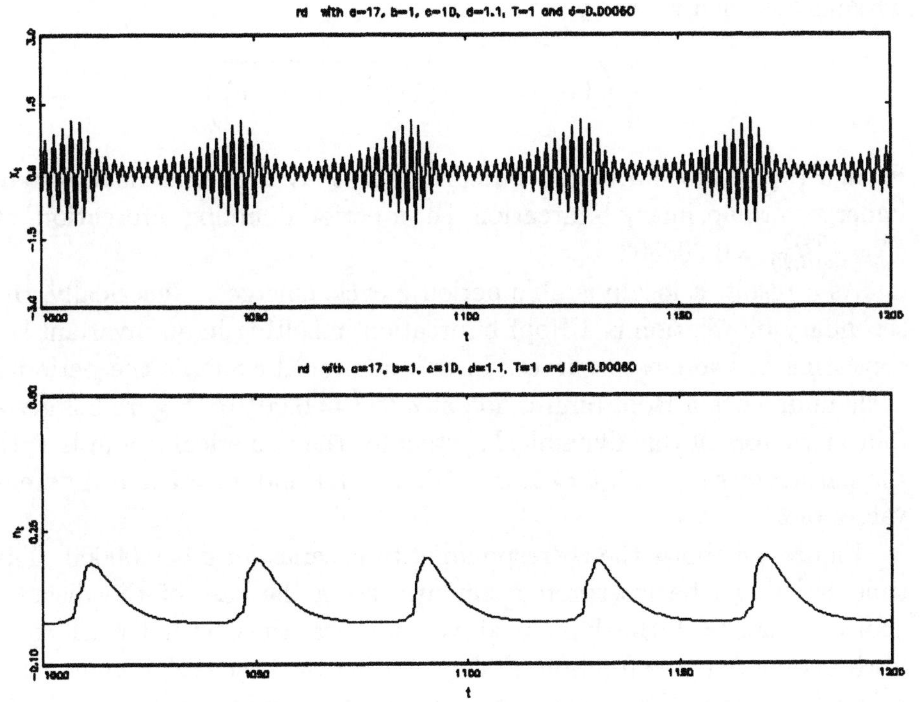

Figure 5.4: Time series of quantities and fraction of rational players for the model with best-reply versus rational players and replicator dynamics with $a = 17$, $b = 1$, $c = 10$, $d = 1.1$, $T = 1$ and $\delta = 0.00060$.

of two closed curves. We conjecture, that a Hopf bifurcation occurs when the period two orbit reaches a point at which profits for the rational firms and for the best-reply firms are equal (that is when $\Pi^r\left(\widehat{X},\widehat{n}\right) - T = \Pi^b\left(\widehat{X},\widehat{n}\right)$, where $\left(\widehat{X},\widehat{n}\right)$ is a point of the period two orbit), since from then on the best-reply strategy is not always more profitable any more, which is a possible explanation of the fact that the fraction of rational agents starts fluctuating.[2] Now let us consider some numerical results obtained by simulations. In these simulations we let $a = 17$, $b = 1$, $c = 10$, $d = 1.1$ and $T = 1$. The equilibrium fraction of the population

[2]The set of point where profits for both rules are equal can easily be computed as follows. We have $\Pi^r - \Pi^{br} = \left(b - \frac{1}{2}d\right)\left(\frac{3b-d}{(2+n)b-d}\right)^2 X^2$ and therefore $\Pi^r - T = \Pi^{br}$ corresponds to points (X, n) which satisfy $X = \pm\frac{(2+n)b-d}{3b-d}\sqrt{\frac{2T}{2b-d}}$.

dynamics is then given by

$$n^* = \frac{1}{2}\left(11\delta + 1 - \sqrt{(11\delta)^2 + (1 - 2\delta)}\right).$$

From Proposition 5.5 it can be concluded that the dynamical system undergoes the primary bifurcation, i.e. a period doubling bifurcation, at $\delta^* = \frac{6859}{801020} \approx 0.008563$.

As a result, a locally stable period 2 cycle emerges. Generically, the secondary bifurcation is a Hopf bifurcation, resulting in an invariant set consisting of two closed curves. In this numerical example the period 2 cycle undergoes a Hopf bifurcation at $\delta^{Hopf} \approx 0.00107$. Figure 5.3 gives four attractors of this dynamical system for the numerical example with the parameters $a = 17$, $b = 1$, $c = 10$, $d = 1.1$ and $T = 1$ and different values of δ.

Figure 5.4 shows the corresponding time series for $\delta = 0.00060$. The time series can be interpreted similarly as in the case of the discrete choice dynamics. If the dynamical system is near the Cournot-Nash equilibrium quantity, the fraction of players using the best-reply behavioural rule will increase. This causes the dynamical system to become unstable and drives the quantities away from the Cournot-Nash equilibrium quantity, making it profitable for the players to use the rational behavioural rule. Consequently, the fraction of rational players will increase and the dynamical system becomes stable again, i.e. quantities return to the Cournot-Nash equilibrium. The main difference between the discrete choice dynamics and the replicator dynamics is that the latter responds less fast to differences in the fitness of the behavioural rules, as can clearly be seen by comparing Figures 5.2 and 5.4. As mentioned in Section 5.2 this is due to the fact that for the replicator dynamics population fractions are partly determined by the old levels of the population fractions, whereas population fractions in the discrete choice model are determined only by the difference in profits between behavioural rules.

5.5 Global Bifurcation Analysis

In the previous section we have seen what happens if evolutionary pressure increases (corresponding to an increase in β for the discrete choice model or a decrease in δ for the replicator dynamics). In particular, simulations suggested that as this evolutionary pressure increases a bifurcation route to strange attractors occurs. In this section we will provide

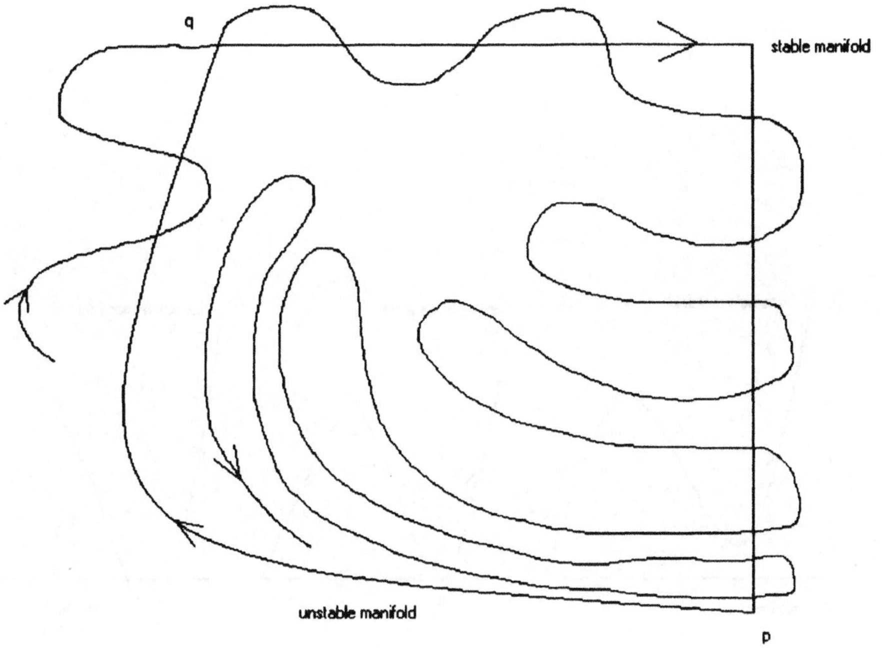

Figure 5.5: Shape of stable and unstable manifolds ($W^s(p)$ and $W^u(p)$) if there is a homoclinic intersection point q .

computer assisted proofs of the existence of these strange attractors by establishing that a homoclinic bifurcation between the stable and unstable manifolds of the unstable equilibrium occurs. We only do this for the replicator dynamics since the model for the discrete choice dynamics is (upon a transformation of variables) equivalent to the cobweb model analyzed in Brock and Hommes (1997) and the existence of strange attractors was already proved there.

5.5.1 Homoclinic Bifurcation Theory

In this section we will briefly describe the theory of homoclinic bifurcations. For a comprehensive treatment of this theory we refer to Palis and Takens (1993). Consider a differentiable two-dimensional map $F_\alpha : \mathbf{R}^2 \to \mathbf{R}^2$, where $\alpha \in \mathbf{R}$ is a parameter. Let p be a saddle fixed point, that is, let the Jacobian of F_α evaluated at p have two real eigen-

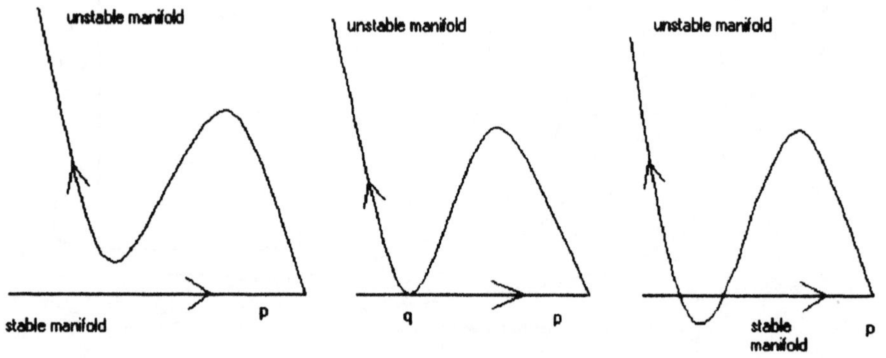

Figure 5.6: Homoclinic bifurcation. a) Stable and unstable manifolds before the bifurcation ($\alpha > \alpha_0$), b) Stable and unstable manifolds at the homoclinic bifurcation ($\alpha = \alpha_0$) and c) Stable and unstable manifolds after the homoclinic bifurcation ($\alpha < \alpha_0$).

values λ_1 and λ_2 where $0 < |\lambda_2| < 1 < |\lambda_1|$.[3] We can now define the *local stable and unstable manifolds* of the equilibrium p as

$$\begin{aligned} W_{loc}^{s}(p) &= \left\{x \in U | F^t(x) \to p \text{ for } t \to +\infty\right\}, \\ W_{loc}^{u}(p) &= \left\{x \in U | F^t(x) \to p \text{ for } t \to -\infty\right\}, \end{aligned}$$

where U is some small neighbourhood of p. These local manifolds are tangent to the corresponding stable and unstable eigenvectors of the linearized dynamical system at the equilibrium (this follows from the *stable manifold theorem* (see for instance Guckenheimer and Holmes (1983), p.

[3]The theory on homoclinic bifurcations can also be applied to a periodic saddle, but for the moment we focus on fixed points of F.

18)). The *global stable and unstable manifolds* are then defined as

$$W^s(p) = \cup_{t\leq 0} F^t(W^s_{loc}(p)) \text{ and } W^u(p) = \cup_{t\geq 0} F^t(W^u_{loc}(p)).$$

Notice that these manifolds are invariant under F, that is, $F(W^s(p)) = W^s(p)$ and $F(W^u(p)) = W^u(p)$. If F is a linear system, these manifolds correspond to the stable and unstable eigenvectors of the linearized system. However, for a nonlinear mapping the stable and unstable manifolds can have a more complicated structure. In fact, they can have intersections, that is, there may exist a point $q \neq p$ with $q \in W^s(p) \cap W^u(p)$. Such a point q is called a *point of homoclinic intersection.* Since the stable and unstable manifold are invariant under F, we then must have that $F^t(q)$, $t = \pm 1, \pm 2, \ldots$, are also points of homoclinic intersection. The sequence of points $\{F^t(q)\}_{t=-\infty}^{t=\infty}$ then is called a *homoclinic orbit.* The existence of such a homoclinic orbit implies a very complicated structure of the unstable and stable manifolds. Since the sequence $F^t(q)$ (by definition) converges to p when $t \to -\infty$, and all these points lie in the stable and unstable manifold the unstable manifold accumulates onto itself infinitely often as $F^t(q)$ approaches p. The same holds for the stable manifold. Figure 5.5 shows the shape that the unstable and stable manifold have if there is a point of homoclinic intersection.

The map F, and therefore also the stable and unstable manifolds, depends upon the parameter α. A *homoclinic bifurcation* is said to occur at $\alpha = \alpha_0$ when for $\alpha > \alpha_0$ there is no intersection between the unstable manifold $W^u(p)$ and the stable manifold $W^s(p)$, for $\alpha = \alpha_0$ there is a point of homoclinic tangency between $W^s(p)$ and $W^u(p)$ and for $\alpha < \alpha_0$ there is a point of transversal homoclinic intersection. Figure 5.6 shows the creation of such a homoclinic bifurcation as α decreases. Such a homoclinic bifurcation implies all kinds of complicated behaviour. First of all, there have to be wild oscillations of the unstable and stable manifolds as pointed out above. Furthermore, it can be shown that if the map F has a transversal homoclinic orbit there exist infinitely many *horseshoes.* This implies the existence of an invariant *Cantor* set Λ containing infinitely many periodic points and an uncountable set of aperiodic points. F then exhibits *sensitive dependence on initial conditions* with respect to Λ. Since the set Λ may have Lebesgue measure zero, the complicated behaviour for this set may only influence the transient behaviour of almost all of the generated orbits of the map F.

Other interesting dynamical phenomena occur in the interval $(\alpha_0 - \varepsilon, \alpha_0 + \varepsilon)$, with $\varepsilon > 0$ small, when the equilibrium is dissipative at the homoclinic bifurcation, that is, if the product of eigenvalues of the Jacobian matrix evaluated at p is smaller than 1 in absolute value. First, this implies existence of Henon-like strange attractors for an open interval of $\alpha-$values with positive Lebesgue measure (Benedicks and Carleson (1991) and Mora and Viana (1993)). Second, there is coexistence of infinitely many stable cycles for a residual set of $\alpha-$values (this is called the Newhouse phenomenon, Newhouse (1974,1979)). Finally, there are cascades of infinitely many period doubling and period halving bifurcations (Yorke and Alligood (1983)). In the following sections we will demonstrate that in our model a homoclinic bifurcation occurs and therefore all these phenomena occur in our evolutionary Cournot duopoly game.

A homoclinic bifurcation in the Cournot model

The first step in the global analysis lies in identifying parts of the stable and unstable manifolds. We know that for values of δ below δ^* (remember that this is the value of δ at which the equilibrium becomes unstable), the equilibrium is a saddle-point since λ_1 is smaller than -1 and λ_2 lies between 0 and 1. The corresponding eigenvectors are given in Appendix 5.8.3. Now recall that our dynamical system (5.14) is given by

$$\begin{aligned} X_{t+1} &= f(X_t, n_t) = -\frac{b(1-n_t)}{(2+n_t)b-d} X_t \qquad (5.6) \\ n_{t+1} &= g(X_t, n_t) \\ &= (1-2\delta)\frac{n_t(\Pi^r(X_t,n_t)-T)}{n_t(\Pi^r(X_t,n_t)-T)+(1-n_t)\Pi^b(X_t,n_t)} + \delta. \end{aligned}$$

We first determine part of the stable manifold. The first picture in Figure 5.7 shows three components of this stable manifold.

The first component (component I) consists of the points with $X = 0$. Clearly, if according to both rules the Cournot-Nash equilibrium quantity is played then neither the best-reply rule nor the rational rule will prescribe a deviation from this Cournot-Nash quantity. Therefore $f(0,n) = 0$, for all n. Furthermore, for $X = 0$, the evolution of the fraction of rational players is governed by

$$n_{t+1} = g(0, n_t) = (1-2\delta)\, n_t \frac{\Pi - T}{\Pi - n_t T} + \delta,$$

where Π are the profits made in equilibrium. $g(0,n)$ is an upward sloping function with a unique equilibrium, at which the slope lies between 0 and 1 (which, of course, is equal to the second eigenvalue λ_2). Therefore the

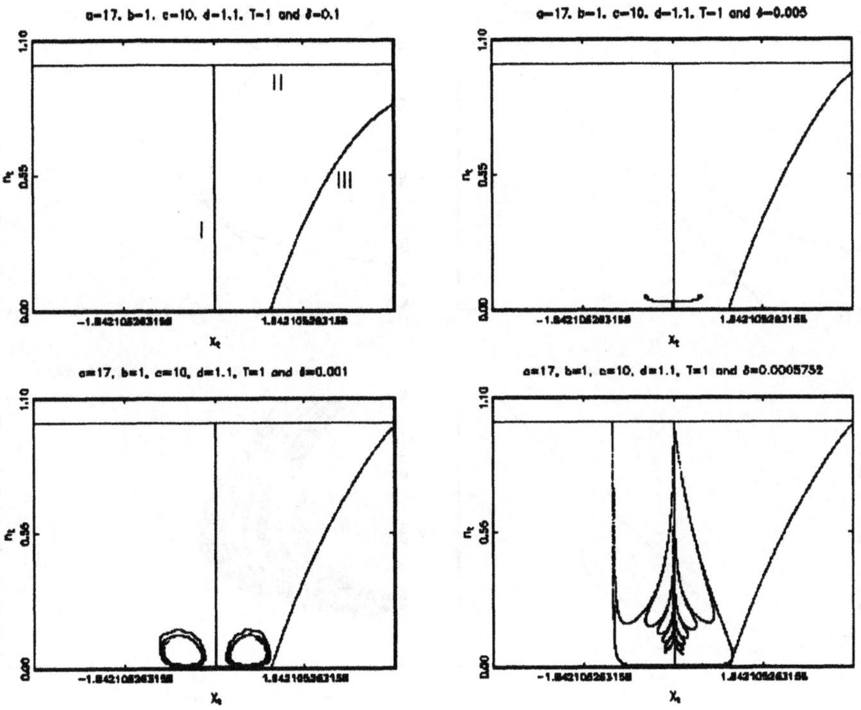

Figure 5.7: Stable and unstable manifolds for the evolutionary model of Cournot competition with replicator dynamics with $a = 17$, $b = 1$, $c = 10$, $d = 1.1$, $T = 1$. a) Three parts of the stable manifold for $\delta = 0.1$, b) Stable and unstable manifolds for $\delta = 0.005$, c) Stable and unstable manifolds for $\delta = 0.001$and d) Stable and unstable manifolds for $\delta = 0.0005752$.

fractions converge to the unique equilibrium n^*. The second component (II) consists of the points where all firms use the rational rule, i.e. where $n = 1$. If all firms are rational, everybody will play the Cournot-Nash equilibrium in the next period. And indeed as can be seen from (5.14) the points with $n = 1$ are mapped onto the point $(0, 1 - \delta)$, which lies in component I. Therefore all points of component II converge to the equilibrium by the same argument as above, so that component II must be part of the stable manifold. We can take this argument one step further by considering points (X, n) which are mapped onto II, i.e. points with $n = 1$, that is, points (X, n) satisfying

$$g(X, n) = 1.$$

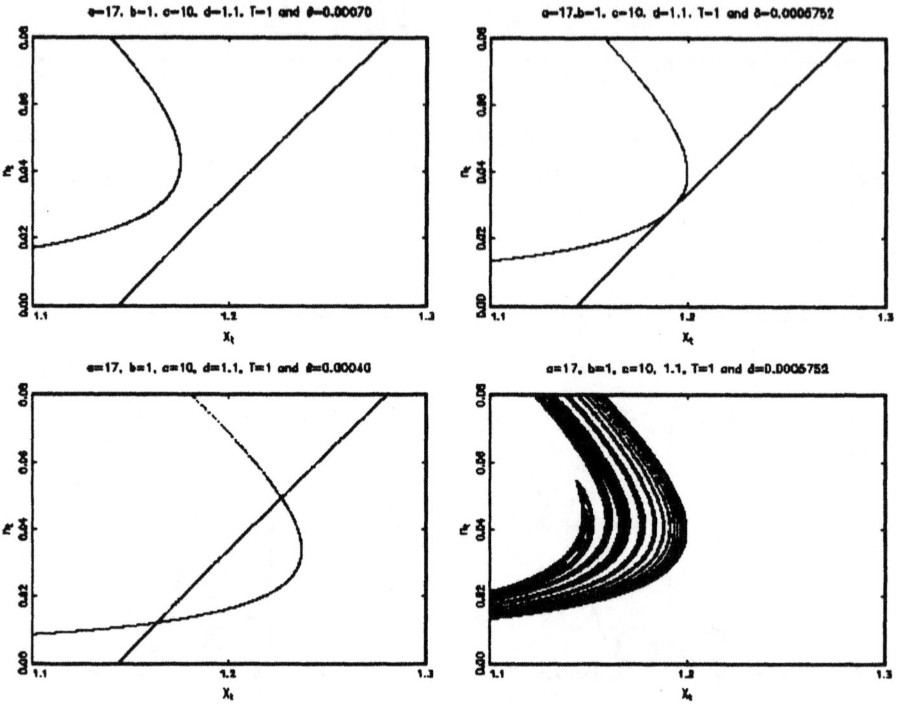

Figure 5.8: Homoclinic bifurcation: a) Stable and unstable manifolds for $\delta = 0.00070$, b) Stable and unstable manifolds for $\delta = 0.0005752$, c) Stable and unstable manifolds for $\delta = 0.00040$ and d) Part of the strange attractor for $\delta = 0.0005752$.

Since $\Pi^r(X, n)$ and $\Pi^b(X, n)$ are quadratic in X we can compute two branches of solutions for this equation. The necessary computations are given in appendix 5.8.4. One of these branches can be ignored, since it has $X < -x^*$ and therefore corresponds to negative production, the other branch corresponds to component III in Figure 5.7. The points in this third component are mapped onto points with $n = 1$, for which the argument given above can again be applied. Notice that this third component varies with the parameters of the model (in particular, it depends upon δ), whereas the first two components are independent of parameters.

We have been able to explicitly compute a part of the stable manifold. We can not do this for the unstable manifold. However, since we know that the local unstable manifold lies tangent to the unstable eigenvector at the equilibrium we can approximate the local unstable manifold by a small piece of the unstable eigenvector. We can then approximate the global unstable manifold by iterating this small eigenvector a number of times. For our numerical example with $a = 17$, $b = 1$, $c = 10$, $d = 1.1$ and $T = 1$, Figure 5.7 shows the stable and unstable manifolds for different values of δ. In the second picture of Figure 5.7 ($\delta = 0.005$) we see that the unstable manifold of the equilibrium approaches the stable period two orbit and in the third picture ($\delta = 0.001$) it spirals to a set of two invariant circles. In the final picture there seems to be a homoclinic tangency between the unstable manifold and component III of the stable manifold of the equilibrium. This happens at $\delta = \delta^c \approx 0.005752$. Figure 5.8 shows an enlargement of a small piece of the stable and unstable manifold. From the first three pictures it is clear that a homoclinic bifurcation occurs at $\delta^c \approx 0.0005752$. For $\delta = 0.00070$ there is no intersection between the stable and the unstable manifold, for $\delta = \delta^c$, there is a tangency and for $\delta = 0.00040$ there are two points of transversal homoclinic intersection. Figure 5.8 is thus numerical evidence of a homoclinic bifurcation as the parameter δ decreases and approaches 0. The fourth picture shows the attractor on this part of the state space. This attractor seems to have a complicated, Cantor-like, structure.

Furthermore, at $\delta = 0.0005752$, we have $|\lambda_1 \times \lambda_2| \approx 0.92235$, hence the equilibrium is dissipative at the homoclinic bifurcation and the results from the previous section apply. Therefore complicated phenomena, such as the existence of Hénon like strange attractors, occur for a set of parameter values in $(\delta^c - \varepsilon, \delta^c + \varepsilon)$ of positive Lebesgue measure.

We now want to point out an important difference between the replicator dynamics and the discrete choice dynamics. Since the discrete choice model implies that fractions always will be strictly between 0 and 1, there is no homoclinic tangency of the stable and unstable manifolds of the equilibrium. To establish complicated behaviour in the cobweb model with discrete choice dynamics (which is equivalent to the Cournot model with discrete choice dynamics) Brock and Hommes (1997) show that (the fourth iterate of) their mapping has a full horseshoe, which implies that there must be homoclinic tangencies between the stable and unstable manifolds of some periodic saddle.

As shown above, for the model with replicator dynamics, we are able to show that a homoclinic bifurcation occurs associated to the saddle point equilibrium. However, we have to deal with the following interpretation problem. If there is a transversal homoclinic intersection, part of the unstable manifold has $n > 1$. Of course, a fraction which becomes

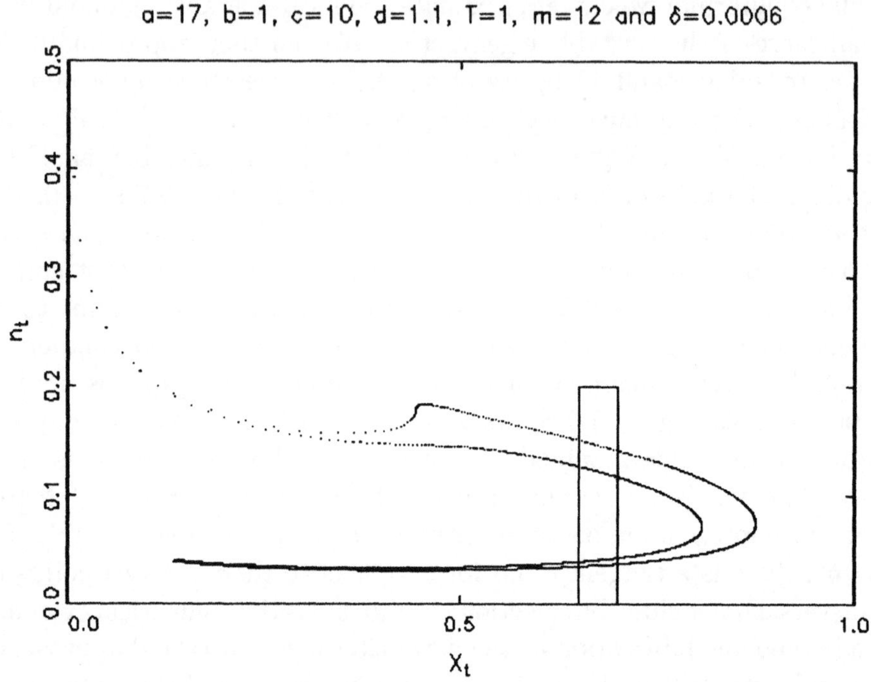

Figure 5.9: Horseshoe in the evolutionary model of Cournot competition: the rectangular box is mapped into the horseshoe-like figure in 12 iterations..

larger than one, has no real meaning in the interpretation of the model. Therefore we would like to show that the complicated phenomena occur already before the homoclinic bifurcation, that is, in the interval $(\delta^c, \delta^c + \varepsilon)$, where everything is well-defined. The fourth picture in Figure 5.3 suggest that this is indeed the case, since there it looks like we have a strange attractor and $\delta = 0.00060 > \delta^c$. The results on homoclinic bifurcation theory support the conjecture that strange attractors occur before the bifurcation value. Finally we mention that the 12'th iterate of the mapping F of our model has a full horseshoe for $\delta = 0.0006$. This horseshoe is shown in Figure 5.9. Therefore we know that there has to be a homoclinic bifurcation between the stable and unstable manifolds of some periodic saddle for some $\delta > \delta^c$.

Sketch of the proof of a homoclinic bifurcation

Most of homoclinic bifurcation theory applies to diffeomorphisms, that is, smooth maps with a smooth inverse. However, our map (5.14) is non-invertible. So we first analyze the special structure of our model. To do this it is helpful to introduce the following three curves (appendix 5.8.4 gives explicit expressions for these curves)

$$\begin{aligned} L^b &= \left\{(X,n)\,|\Pi^b(X,n)=0\right\}, L^r = \{(X,n)\,|\Pi^r(X,n)=T\} \text{ and} \\ L^n &= \left\{(X,n)\,|n\left(\Pi^r(X,n)-T\right)+(1-n)\,\Pi^b(X,n)=0\right\}. \end{aligned}$$

These curves are shown in Figure 5.10. L^b and L^r are curves that give combinations of X and n at which, respectively, the profit for best-reply agents and the net profit for rational agents are equal to 0. The curve L^n gives the points (X,n) at which the numerator in $g(X,n)$ vanishes. Therefore the map (5.14) is not well-defined on L^n. Notice that these curves do not depend on the amount of deterministic noise δ. It can be easily seen that the curve L^r lies to the right of L^b. From its definition it follows that L^n must lie between L^b and L^r. L^b and L^n intersect the $n-$axis in the point $\left(X^f,0\right)$, where X^f is such that $\Pi^b\left(X^f,0\right)=0$. In this point the map contains an element 0/0, and such a point is called a *focal* point. Focal points can play an important role in the global dynamics of noninvertible dynamical systems, see for example Bischi, Gardini and Mira (1999).

Our next step is to see when the map F has a smooth inverse. In order to do this we analyze the determinant of the Jacobian matrix. If this determinant is equal to 0, the system is not invertible at that point. It is impossible to determine analytically when the determinant disappears, however we know that it vanishes in the focal point $\left(X^f,0\right)$, when $n=1$ and on the curve L^r. Figure 5.10 shows the curves L^b, L^r and L^n and the curves along which the determinant of F vanishes (L^r and $L^{\det}$). We have checked that the only point where the determinant vanishes together with the curve L^n is the focal point. This implies that to the left of L^n the dynamical system is locally a diffeomorphism (except at the focal point).

Now we focus attention on the component of the stable manifold, that is mapped into points with $n=1$. This component is

$$L_\delta^{III}=\{(X,n)\,|g(X,n)=1\}.$$

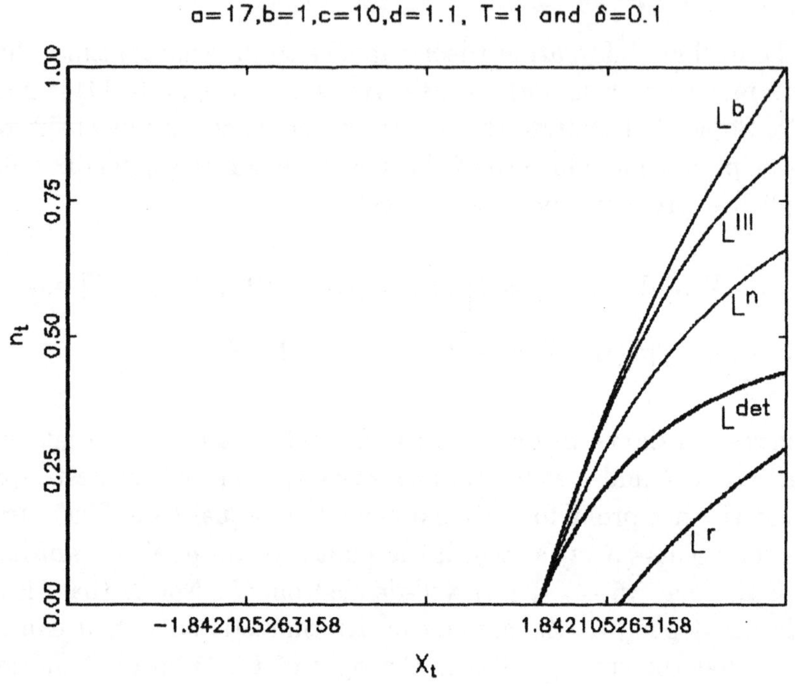

Figure 5.10: Some important curves: L^b and L^r are the curves along which net profits for the best-reply and rational rule are zero, respectively. L^n is the curve along which the denominator of $g(X,n)$ vanishes, L^{det} is the curve along which the determinant of the Jacobian matrix vanishes and L^{III} is the third component of the stable manifold of the equilibrium.

Now consider

$$g(X,n) = (1-2\delta)\frac{n(\Pi^r(X,n)-T)}{n(\Pi^r(X,n)-T)+(1-n_t)\Pi^b(X,n)}+\delta = 1$$

and notice that for this equality to hold we must have $\Pi^b(X,n) < 0$, $\Pi^r(X,n) > 0$ and $n(\Pi^r(X,n)-T)+(1-n_t)\Pi^b(X,n) > 0$. In particular for $\delta = 0$, we have $L_0^{III} = L^b$. As δ increases profits associated to the best-reply rule have to decrease even further and it can be easily seen that as δ approaches $\frac{1}{2}$, L_δ^{III} approaches L^n. Clearly, for $0 < \delta < \frac{1}{2}$, L_δ^{III} lies between L^b and L^n and as δ increases it shifts from L^b to L^n. Furthermore, the dynamical system is locally a diffeomorphism in the neighbourhood of L_δ^{III}.

We want to establish that a homoclinic bifurcation occurs. Now first we briefly consider the case $\delta = 0$. Notice that this corresponds to the replicator dynamics without deterministic noise. The system then becomes

$$\begin{aligned} X_{t+1} &= -\frac{b(1-n_t)}{(2+n_t)b-d}X_t \\ n_{t+1} &= \frac{n_t(\Pi^r(X_t,n_t)-T)}{n_t(\Pi^r(X_t,n_t)-T)+(1-n_t)\Pi^b(X_t,n_t)}. \end{aligned} \tag{5.15}$$

If $T > 0$ we have two equilibria: $(X^*, n^*) = (0,0)$ and $(X^{**}, n^{**}) = (0,1)$.[4] Hence, in an equilibrium the whole population is playing the same rule. Clearly the equilibrium where all agents play the rational behavioural rule is an unstable equilibrium since a deviation to the best-reply strategy pays off since then information costs T do not have to be incurred.[5] On the other hand, the equilibrium where all agents use the best-reply behavioural rule is unstable when the best-reply dynamics are unstable. We are primarily interested in the equilibrium $(X^*, n^*) = (0,0)$, since the other equilibrium disappears when $\delta > 0$. From (5.15) it follows that the stable manifold of this equilibrium is the $n-$axis, the unstable manifold is the $X-$axis. Notice that the unstable manifold contains the focal point $(X^f, 0)$. We know that the unstable manifold varies continuously with δ. We now want to show that if δ is a small but positive number, there is a point of homoclinic intersection. For δ small but positive we know that part of the stable manifold lies close to L^b and the unstable manifold lies close to the $X-$axis. Furthermore, we know that for δ large, say below but close to δ^*, there is no intersection between the stable and unstable manifolds of the equilibrium of the system, see for example the second picture in Figure 5.7. The unstable manifold then lies above and to the left of L^b and L_δ^{III} lies to the right of L^b. Therefore, by a continuity argument, we can conclude that there has to exist a δ^c between 0 and δ^*, for which there is a homoclinic tangency between the stable and unstable manifolds of the equilibrium.

We now show that indeed, for δ sufficiently small there is an intersection between the stable and unstable manifolds. In order to do this

[4] If $T = 0$ there is a continuum of equilibria: all states with $X = 0$ are equilibria, since profits for both rules are equal.

[5] Using terminology from evolutionary game theory, we say that this equilibrium is not *evolutionary stable*, since it can be *invaded* by the best-reply strategy, which, in the neighbourhood of this equilibrium does better against the rational strategy than the rational strategy does against itself.

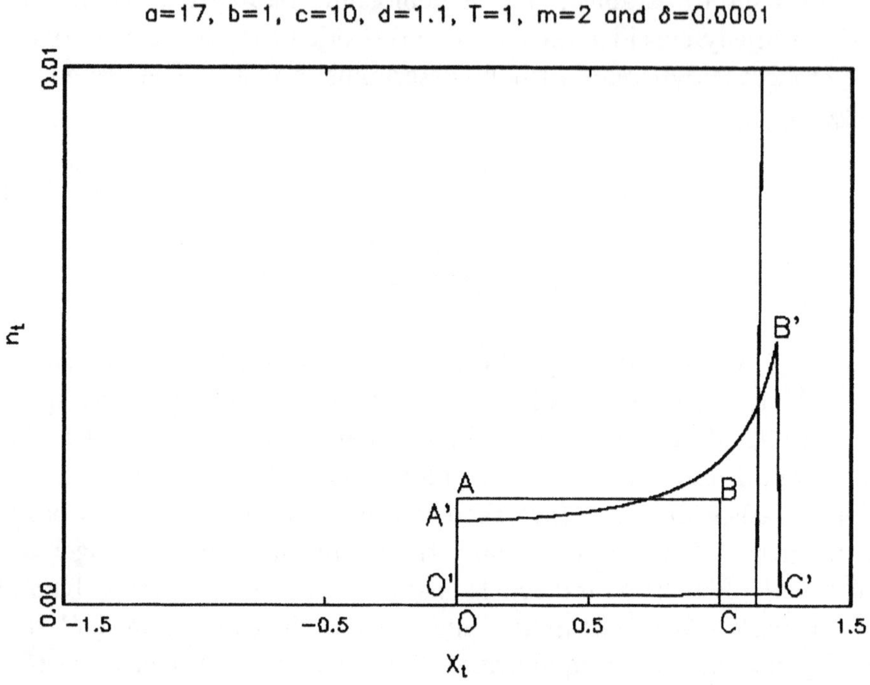

Figure 5.11: Graphical sketch of the proof of homoclinic intersection.

we construct a box $OABCO$ with $O = (0,0)$, $A = (0, n_0)$, $B = (X_0, n_0)$ and $C = (X_0, 0)$, where $0 < X_0 < X^f$ and $0 < n^* < n_0$, see Figure 5.11. We know that for δ small enough, by the continuity argument made above, this box contains a piece of the unstable manifold and this unstable manifold leaves the box via the line segment BC. We then have to show that the second iterate of this box, which we shall call $O'A'B'C'O' = F^2(OABCO)$ has a transversal intersection with the stable manifold. Now consider our numerical example with $a = 17$, $b = 1$, $c = 10$, $d = 1.1$ and $T = 1$. Furthermore take $\delta = 0.0001$, $X_0 = 1$ and $n_0 = 0.002$. For these parameter values we have $n^* \approx 0.00061$ and $X^f \approx 1.1434$ and therefore X_0 and n_0 satisfy the necessary restrictions. We checked that B' and C' lie to the right of $L^{III}_{0.0001}$, which implies that also the unstable manifold has to cross $L^{III}_{0.0001}$. Therefore, for $\delta = 0.0001$ we have a transversal intersection between the stable and unstable manifolds of the equilibrium. Furthermore, since Figure 5.7 shows that there is not such a intersection for $\delta = 0.005$, by continuity there must be a value $\delta^c \in (0.005, 0.0001)$ such that for δ^c the unstable manifold and the stable manifold have a homoclinic tangency. Since the map is locally a diffeomorphism around points at $L^{III}_{0.0001}$ we can apply the results on homoclinic bifurcations and conclude that all kinds of strange behaviour referred to in Section 5.5.1, occur in our model.

5.6 Imitators versus Best-Reply Players

As a second example we confront best-reply players with imitators. Just like the model in Section 5.3, best-reply players respond optimally to the average industry output from the previous period. Imitators copy the quantity produced by the best-reply players in the previous period. As a consequence, the imitation rule is relatively cheap compared to the best-reply rule since it does not require any knowledge of the underlying Cournot model. Now let x_t denote the quantity supplied by the best-reply players, z_t the quantity supplied by the imitators, and n_t the fraction of the population using the best-reply behavioural rule in period t. Consequently, the complete dynamics of the model are given by

$$\begin{aligned} x_{t+1} &= R\left(nx_t + (1-n)\, z_t\right) \\ z_{t+1} &= x_t \\ n_{t+1} &= h\left(x_t, z_t, n_t\right). \end{aligned} \tag{5.16}$$

Note that this is a 3-dimensional dynamical system which dimension cannot be reduced. Furthermore, the Cournot-Nash equilibrium quantity x^* is not the unique equilibrium quantity of the imitation rule. In fact, any quantity is an equilibrium of the imitation rule. The Cournot-Nash equilibrium quantity is, however, still the unique equilibrium quantity of the complete dynamical system. For this reason we do not have to worry about Assumption 5.1 not being satisfied. Before specifying the population dynamics $h\left(x_t, z_t, n_t\right)$ we have the following result with respect to the dynamical system (5.16).

Lemma 5.3 *Consider the dynamical system (5.16). The eigenvalues of the Jacobian matrix of the linearized system, evaluated at the equilibrium (x^*, x^*, n^*), are*

$$\lambda_{1,2} = \frac{1}{2} n^* R'\left(x^*\right) \pm \frac{1}{2}\sqrt{\left(n^* R'\left(x^*\right)\right)^2 + 4\left(1-n^*\right) R'\left(x^*\right)}$$

and $\lambda_3 = \left.\frac{\partial h(x_t, z_t, n_t)}{\partial n_t}\right|_{(x^*, x^*, n^*)}$.

Proof. It can easily be shown that the Jacobian matrix, evaluated at the equilibrium (x^*, x^*, n^*), is given by

$$\mathbf{J} = \begin{pmatrix} n^* R'(x^*) & (1-n^*) R'(x^*) & 0 \\ 1 & 0 & 0 \\ J_{31} & J_{32} & \left.\frac{\partial h(x_t,z_t,n_t)}{\partial n_t}\right|_{(x^*,x^*,n^*)} \end{pmatrix}.$$

The corresponding eigenvalues are exactly as stated in Lemma 5.3. ■

5.6.1 Discrete Choice Model

First consider the case where population fractions are updated according to the discrete choice model. The model with imitators and best-reply players, as represented by the dynamical system (5.16), then becomes

$$\begin{aligned} x_{t+1} &= R(nx_t + (1-n) z_t) \\ z_{t+1} &= x_t \\ n_{t+1} &= h(x_t, z_t, n_t) \\ &= \frac{\exp\left[\beta\left(\Pi^b(x_t, z_t, n_t) - T\right)\right]}{\exp\left[\beta\left(\Pi^b(x_t, z_t, n_t) - T\right)\right] + \exp\left[\beta \Pi^i(x_t, z_t, n_t)\right]}, \end{aligned} \qquad (5.17)$$

where $\Pi^b(x_t, z_t, n_t)$ and $\Pi^i(x_t, z_t, n_t)$ are the average profit of the best-reply players and the imitators in period t, respectively. We will study what happens to the dynamics of (5.17) as the intensity of choice β increases.

Local Stability Analysis

We have the following result with respect to the local stability of the equilibrium of the dynamical system (5.17).

Proposition 5.6 *Consider the dynamical system (5.17). Let* $-8 < R'(x^*) < -1$. *The equilibrium of (5.17) is* (x^*, x^*, n^*), *where* x^* *is the Cournot-Nash equilibrium quantity and* $n^* = \frac{1}{1+\exp[\beta T]}$. *Then given* T *there exists a value* β^* *of* β *such that the equilibrium* (x^*, x^*, n^*) *is locally stable for all* $\beta < \beta^*$ *and unstable for all* $\beta > \beta^*$. *Moreover,*

$$\beta^* = \frac{1}{T} \ln\left[\frac{1}{-R'(x^*) - 1}\right].$$

Proof. Straightforward calculations show that (x^*, x^*, n^*) is the equilibrium of (5.17). First, it can easily be checked that the eigenvalue $\lambda_3 = 0$. Second, the other two eigenvalues λ_1 and λ_2 are complex conjugates if and only if

$$\left(n^* R'(x^*)\right)^2 + 4(1-n^*) R'(x^*) < 0.$$

Rearranging terms shows that this condition is equivalent to

$$-R'(x^*) < 4\frac{1-n^*}{(n^*)^2}.$$

Because of the information costs $T > 0$ we know that $n^* \in \left(0, \frac{1}{2}\right]$. Consequently, the above condition is always satisfied if $R'(x^*) > -8$. The equilibrium (x^*, x^*, n^*) loses its stability when the eigenvalues $\lambda_{1,2}$ are on the unit circle, i.e.

$$\lambda_1 \times \lambda_2 = -(1-n^*) R'(x^*) = 1 \Leftrightarrow n^* = \frac{1+R'(x^*)}{R'(x^*)}.$$

Substituting $n^* = \frac{1}{1+\exp[\beta T]}$ and solving for β gives the condition as stated in Proposition 5.6. ∎

Note that the equilibrium of (5.17) is locally stable for all values of the intensity of choice $\beta > 0$ in case $-1 \leq R'(x^*) \leq 0$.

A Bifurcation Scenario

In case we consider a linear inverse demand function and quadratic cost functions, the dynamical system (5.17) rewritten in deviations becomes

$$\begin{aligned} X_{t+1} &= -\frac{b}{2b-d}\left(n_t X_t + (1-n_t) Z_t\right) \\ Z_{t+1} &= X_t \\ n_{t+1} &= \frac{1}{1+\exp\left[\beta \Pi^d (X_t, Z_t, n_t)\right]}, \end{aligned} \tag{5.18}$$

where $X_t = x_t - x^*$ and $Z_t = z_t - x^*$ and where

$$\begin{aligned} \Pi^d(X_t, Z_t, n_t) &= \Pi^i(x_t, z_t, n_t) - \Pi^b(x_t, z_t, n_t) + T \\ &= T - b(Z_t - X_t)\left(n_t X_t + (1-n_t) Z_t\right) \\ &\quad + \left(\frac{1}{2}d - b\right)\left(Z_t^2 - X_t^2\right) \end{aligned}$$

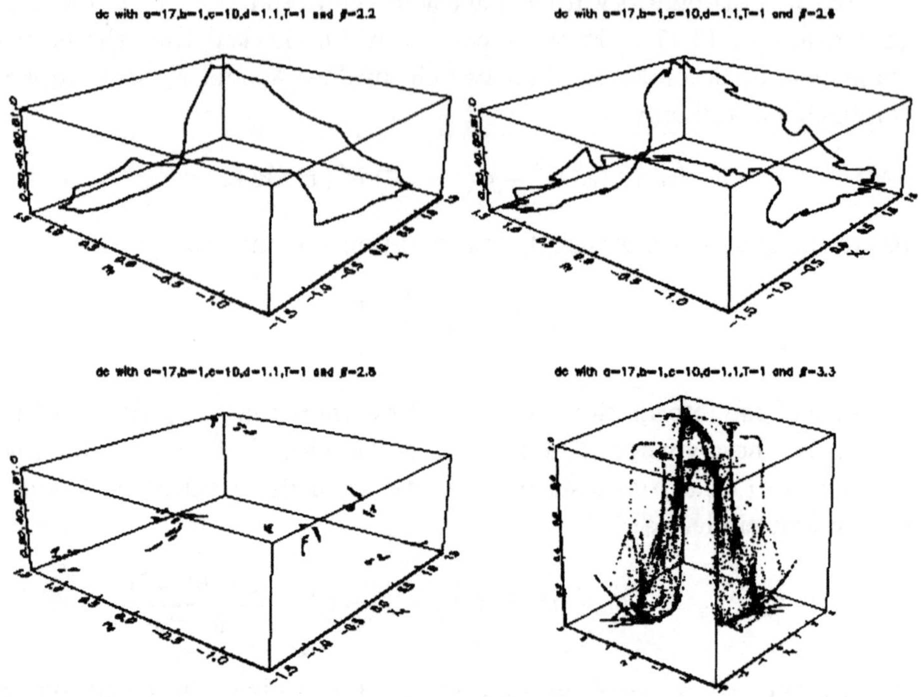

Figure 5.12: Attractors for the model with imitators versus best-reply players and discrete choice dynamics with $a = 17$, $b = 1$, $c = 10$, $d = 1.1$, $T = 1$ and a) $\beta = 2.2$, b) $\beta = 2.6$, c) $\beta = 2.8$ and d) $\beta = 3.3$.

are deviations from the Cournot-Nash equilibrium quantity x^*. As indicated by Proposition 5.6 a Hopf bifurcation occurs at $\beta^* = \frac{1}{T}\ln\left[\frac{2b-d}{d-b}\right]$. In fact, the equilibrium of the dynamical system becomes unstable and an invariant closed curve emerges. Because the primary bifurcation is a Hopf bifurcation we cannot use the symmetry of the dynamical system (5.18) to determine the secondary bifurcation analytically. The invariant set resulting after the Hopf bifurcation consists of infinitely many points which cannot be written down explicitly. Consequently, it is not possible to determine the Jacobian matrix. As β increases further the closed curve breaks up and a strange attractor is created. Figure 5.12 gives pictures of the attractors for parameter values $a = 17$, $b = 1$, $c = 10$, $d = 1.1$, $T = 1$, and for different values of β. For this numerical example the primary bifurcation occurs at $\beta^* = 2\ln 3$.

5.6.2 Replicator Dynamics

Consider the model with imitators versus best-reply players and the population dynamics described by the replicator dynamics with deterministic noise. This results in a specification of the dynamical system (5.16) given by

$$\begin{aligned} x_{t+1} &= R\left(nx_t + (1-n)\,z_t\right) \\ z_{t+1} &= x_t \\ n_{t+1} &= \frac{n_t\,(1-2\delta)\left(\Pi^b\left(X_t, Z_t, n_t\right) - T\right)}{n_t\left(\Pi^b\left(X_t, Z_t, n_t\right) - T\right) + (1-n_t)\,\Pi^i\left(X_t, Z_t, n_t\right)} + \delta \end{aligned} \tag{5.19}$$

In this section we study what happens to the dynamical behaviour of (5.19) as the deterministic noise δ approaches zero.

Local Stability Analysis

The local stability analysis of the equilibrium of the dynamical system (5.19) is summarized in Proposition 5.7.

Proposition 5.7 *Consider the dynamical system (5.19). Let* $0 < \delta < \frac{1}{2}$, $T > 0$, *and* $-8 < R'(x^*) < -1$. *The equilibrium of (5.19) is* (x^*, x^*, n^*), *where* x^* *is the Cournot-Nash equilibrium quantity and*

$$n^* = \frac{1}{2T}\left(\delta\,(2\Pi^* - T) + T - \sqrt{\delta^2\,(2\Pi^* - T)^2 + (1-2\delta)\,T^2}\right).$$

Furthermore, the equilibrium (x^*, x^*, n^*) *is locally stable (unstable) for* $\delta > \delta^*$ $(\delta < \delta^*)$, *where*

$$\delta^* = \frac{(R'(x^*) + 1)\,T}{R'(x^*)\left((R'(x^*) + 2)\,(\alpha + \Pi^*) - (R'(x^*) + 1)\,T\right)}.$$

Proof. Straightforward calculations show that (x^*, x^*, n^*) is the equilibrium of (5.19). It can easily be checked that the equilibrium (x^*, x^*, n^*) loses stability when the eigenvalues λ_1 and λ_2, which are complex conjugates, cross the unit circle. This happens when

$$n^* = \frac{1 + R'(x^*)}{R'(x^*)}.$$

Substituting the expression for n^* and solving for δ gives the condition as stated in Proposition 5.7. ■

Note that the equilibrium (x^*, x^*, n^*) of (5.19) is locally stable for all values of the deterministic noise δ in case $-1 \leq R'(x^*) \leq 0$.

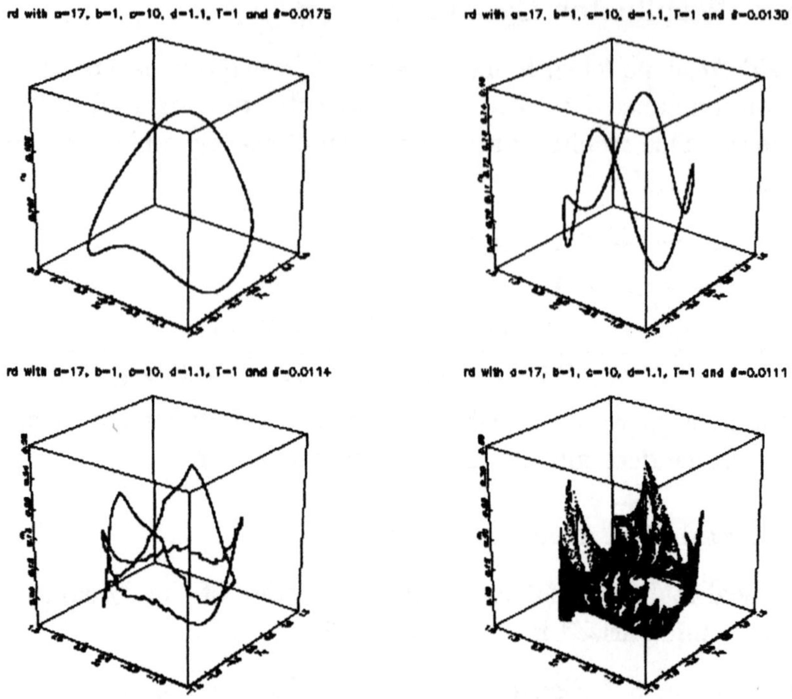

Figure 5.13: Attractors for the model with imitators versus best-reply players and replicator dynamics with $a = 17$, $b = 1$, $c = 10$, $d = 1.1$, $T = 1$ and a) $\delta = 0.0175$, b) $\delta = 0.0130$, c) $\delta = 0.0114$ and d) $\delta = 0.0111$.

A Bifurcation Scenario

We now present some simulation results for the dynamical system (5.19). Let the parameters $a = 17$, $b = 1$, $c = 10$, $d = 1.1$ and $T = 1$. As indicated by Proposition 5.7 a Hopf bifurcation occurs at $\delta^* = \frac{9}{490}$. Figure 5.13 gives attractors for the numerical example with the above parameter values and different values of the deterministic noise δ.

5.7 Concluding Remarks

In this chapter we have considered an evolutionary game theoretic model dealing with the selection of behavioural rules in a Cournot duopoly game. Contrary to most evolutionary game theory, which concentrates on the selection of equilibrium actions in matrix games, we have focussed on a setting in which different types of behavioural rules are selected. The population dynamics representing the evolutionary selection of behavioural rules is coupled with the quantity dynamics arising from the interplay between these different behavioural rules. Once again, note that the results in this chapter can easily be generalized to Cournot oligopoly games. An alternative interpretation of the model presented here is that it deals with a single market with two firms. At each time period the identical firms are characterized by a mixed strategy over the K behavioural rules. In fact, the population fractions should be interpreted as the probabilities with which the two firms choose a certain behavioural rule. When both firms have chosen a behavioural rule they play the Cournot game. Based upon the profits realized in that Cournot game the probabilities with which the firms choose a behavioural rule in the next period will be adjusted. Clearly, this model is mathematically identical to the one discussed in this chapter.

The behaviour of the resulting dynamical system depends critically on the stability of the quantity dynamics. First, in case the cheapest behavioural rule is stable, the evolutionary process converges to a situation where most agents use this behavioural rule and produce quantities equal to the Cournot-Nash equilibrium quantity. Second, if the cheapest behavioural rule is unstable, complicated dynamical behaviour may occur. In particular, high evolutionary pressure or a small noise rate with respect to the choice of behavioural rules leads to highly irregular quantity dynamics converging to a strange attractor.

Two typical examples have been analyzed to illustrate the complicated dynamical behaviour in detail. The fractions of the population using each of the two behavioural rules were determined by the discrete choice dynamics or the replicator dynamics with deterministic noise. In case the population dynamics is described by the discrete choice model, we found that if the simple behavioural rule is unstable and natural selection with respect to the behavioural rules is strong enough, strange dynamical behaviour may occur. In case we use the replicator dynamics to describe the updating of the population fractions, we observe complicated dynamical behaviour if the simple rule is unstable and deterministic noise is small enough. Note that the nonlinearity causing this erratic behaviour arises from the interplay between population dynamics and quantity dynamics and not from a particular nonlinear specification of the underlying Cournot duopoly model.

For the case of best-reply players against rational players we have shown that the complicated behaviour of the Cournot model with the replicator dynamics is due to a homoclinic tangency between the stable and unstable manifolds of the unstable equilibrium. Simulations suggest that this complicated behaviour also is a feature of the evolutionary Cournot model when we focus on other behavioural rules, and that therefore this existence of strange attractors is a robust feature of these kinds of models.

The model presented in this chapter can be extended in a number of interesting directions. The only equilibrium in the model discussed in this chapter is the unique Cournot-Nash equilibrium. The model could be used to study a Cournot game with multiple equilibria, to address the issue of equilibrium selection. Furthermore it would be interesting to investigate a model with behavioural rules that would introduce the possibility of coordination on a 'non-Nash'-equilibrium, such as the joint profit maximum (the cartel solution) or the 'Walrasian' equilibrium (where quantities are chosen such that marginal costs equal the price). Another interesting line of research would be to change the matching structure. We have focused on a global interaction scheme. It might be worthwhile to look at some kind of local interaction structure, where firms are only matched with their direct neighbours. It is interesting to see whether the dynamical features of the global interaction model would be preserved in such a local interaction framework.

Finally, we want to refer to Cox and Walker (1998), who report on some laboratory experiments with a Cournot duopoly model with random matching and exactly the same linear-quadratic specification as in our model. They find that in general the subjects do not converge to the Cournot-Nash equilibrium if the best-reply dynamics are unstable. This seems to corroborate the model of this chapter.

5.8 Appendix

5.8.1 Derivation of Profit Functions

Consider the symmetric Cournot duopoly game with a linear demand function and quadratic cost functions, as specified in Section 5.2.1. In that case, the profit of an individual supplying quantity x_1, when he is matched with an individual supplying x_2, is given by

$$\begin{aligned}\pi(x_1, x_2) &= P(x_1, x_2)\, x_1 - c(x_1) = (a - b(x_1 + x_2))\, x_1 - cx_1 + \frac{d}{2}x_1^2 \\ &= (a - c - bx_2)\, x_1 + \left(\frac{1}{2}d - b\right) x_1^2.\end{aligned}$$

Now suppose the individuals in the population can choose between the behavioural rules H_1 and H_2, which prescribe quantities x_1 and x_2, respectively. Let n denote the fraction of the population supplying x_1. Then, in the case of random matching between individuals, the average profit of someone supplying x_1 can be written as

$$\begin{aligned}\Pi^1(x_1,x_2,n) &= n\pi(x_1,x_1)+(1-n)\pi(x_1,x_2)\\ &= (a-c)x_1+\left(\frac{1}{2}d-b\right)x_1^2-b(nx_1+(1-n)x_2)x_1.\end{aligned}$$

Rewritten in terms of deviations $X_1=x_1-x^*$ and $X_2=x_2-x^*$ from the Cournot-Nash equilibrium quantity x^*, the above expression becomes

$$\begin{aligned}\Pi^1(X_1,X_2,n) &= \left(b-\frac{1}{2}d\right)\left(\frac{a-c}{3b-d}\right)^2+\left(\frac{1}{2}d-b\right)X_1^2\\ &\quad -b(nX_1+(1-n)X_2)\left(X_1+\frac{a-c}{3b-d}\right).\end{aligned}$$

Using a similar argument we can find the average profit of someone supplying x_2. In fact, this profit is given by

$$\begin{aligned}\Pi^2(X_1,X_2,n) &= \left(b-\frac{1}{2}d\right)\left(\frac{a-c}{3b-d}\right)^2+\left(\frac{d}{2}-b\right)X_2^2\\ &\quad -b(nX_1+(1-n)X_2)\left(X_2+\frac{a-c}{3b-d}\right).\end{aligned}$$

Note that only the difference between the average profits $\Pi^1(X_1,X_2,n)$ and $\Pi^2(X_1,X_2,n)$ matters for the discrete choice dynamics. Therefore, an important quantity is

$$\begin{aligned}\Pi^1(X_1,X_2,n)-\Pi^2(X_1,X_2,n) &= \left(\frac{1}{2}d-b\right)(X_1^2-X_2^2)\\ &\quad +b(nX_1+(1-n)X_2)(X_1-X_2).\end{aligned}$$

5.8.2 Equivalence of the Cournot and Cobweb Model

In this appendix we show that the Cournot duopoly model discussed in Section 5.4.1 is mathematically equivalent to the cobweb model with naive versus rational expectations introduced by Brock and Hommes (1997).

Consider the Cournot model with rational players and best-reply players. In case the discrete choice model describes the population dynamics, this Cournot model is represented by the dynamical system (5.12), that is

$$\begin{aligned} X_{t+1} &= -\frac{b\,(1-n_t)}{(2+n_t)\,b-d}X_t \\ n_{t+1} &= \frac{1}{1+\exp\left[\beta\left(T-\left(b-\frac{d}{2}\right)\left(\frac{3b-d}{(2+n_t)b-d}\right)^2 X_t^2\right)\right]}. \end{aligned}$$

The corresponding cobweb model of Brock and Hommes (1997), with a linear demand and supply function, can be represented by

$$\begin{aligned} p_{t+1} &= -\frac{\widehat{b}\,(1-m_t)}{2B+\widehat{b}\,(1+m_t)}p_t \\ m_{t+1} &= \tanh\left(\frac{\beta}{2}\left[\frac{\widehat{b}}{2}\left(\frac{\widehat{b}\,(1-m_t)}{2B+\widehat{b}\,(1+m_t)}+1\right)^2 p_t^2\right]-T\right), \end{aligned}$$

where $\widehat{b}$ and B are the marginal demand and supply parameters, p_t is the market clearing price in period t, and $m_t = n_{1,t} - n_{2,t} = 2n_t - 1$ is the difference in the population fractions. This means that if $m_t = 1$ all agents are rational and if $m_t = -1$ all agents use naive expectations. Rewriting their dynamical system in terms of n_t gives

$$\begin{aligned} p_{t+1} &= -\frac{\widehat{b}\,(1-n_t)}{B+\widehat{b}n_t}p_t \\ n_{t+1} &= \frac{1}{1+\exp\left[\beta\left(T-\frac{\widehat{b}}{2}\left(\frac{\widehat{b}+B}{B+\widehat{b}n_t}\right)^2 p_t^2\right)\right]}. \end{aligned}$$

Clearly, if we take the parameters $\widehat{b} = b$ and $B = 2b - d$ their dynamical system becomes

$$\begin{aligned} p_{t+1} &= -\frac{b\,(1-n_t)}{(2+n_t)\,b-d}p_t \\ n_{t+1} &= \frac{1}{1+\exp\left[\beta\left(T-\frac{b}{2}\left(\frac{3b-d}{(2+n_t)b-d}\right)^2 p_t^2\right)\right]}, \end{aligned}$$

which is equivalent to (5.12) up to the following transformation of variables

$$X_t = \sqrt{\frac{b}{2b-d}} p_t.$$

5.8.3 The Jacobian matrix

In this appendix we derive the Jacobian matrix. The dynamical system $(X_{t+1}, n_{t+1})' = (f(X_t, n_t), g(X_t, n_t))'$ is defined by

$$\begin{aligned} f(X,n) &= -\frac{b(1-n)}{(2+n)b-d} X \\ g(X,n) &= (1-2\delta) \frac{n(\Pi^r(X,n)-T)}{n(\Pi^r(X,n)-T)+(1-n)\Pi^b(X,n)} + \delta, \end{aligned}$$

where

$$\begin{aligned} \Pi^b(X,n) &= \left(b-\frac{1}{2}d\right)\left(\frac{a-c}{3b-d}\right)^2 \\ &\quad -b(1-n)\left(\frac{a-c}{3b-d}\right)\left(\frac{2b-d}{(2+n)b-d}\right)X \\ &\quad -\frac{2b-d}{2}\left(\frac{(4-n)b-d}{(2+n)b-d}\right)X^2 \end{aligned}$$

and

$$\begin{aligned} \Pi^r(X,n) &= \left(b-\frac{1}{2}d\right)\left(\frac{a-c}{3b-d}\right)^2 \\ &\quad -b(1-n)\left(\frac{a-c}{3b-d}\right)\left(\frac{2b-d}{(2+n)b-d}\right)X \\ &\quad +\left(b-\frac{1}{2}d\right)\left(\frac{b(1-n)}{(2+n)b-d}\right)^2 X^2 \\ &= \left(b-\frac{1}{2}d\right)\left(\frac{b(1-n_t)}{(2+n)b-d}X-\frac{a-c}{3b-d}\right)^2. \end{aligned}$$

Straightforward computations give

$$f_X = -\frac{b(1-n)}{(2+n)b-d}, f_n = \frac{b(3b-d)}{((2+n)b-d)^2}X$$

and

$$g_X = (1-2\delta)\frac{n(1-n)\left[\Pi^b\Pi^r_X - (\Pi^r - T)\Pi^b_X\right]}{\left(n(\Pi^r - T) + (1-n)\Pi^b\right)^2}.$$

Finally

$$g_n = (1-2\delta)\frac{\Pi^b(\Pi^r - T) + n(1-n)\left(\Pi^b\Pi^r_n - (\Pi^r - T)\Pi^b_n\right)}{\left(n(\Pi^r - T) + (1-n)\Pi^b\right)^2}.$$

The Jacobian matrix is given by

$$J = \begin{pmatrix} f_X & f_n \\ g_X & g_n \end{pmatrix}.$$

If $X = 0$ (everybody playing the Cournot-Nash equilibrium) we have

$$\Pi^b(0,n) = \Pi^r(0,n) \equiv \Pi,$$

furthermore, from the profit functions it follows that

$$\begin{aligned}
\Pi^r_X &= \frac{b(1-n)(2b-d)}{(2+n)b-d}\left(\frac{b(1-n)}{(2+n)b-d}X - \frac{a-c}{3b-d}\right), \\
\Pi^b_X &= -\frac{2b-d}{(2+n)b-d}\left(b(1-n)\left(\frac{a-c}{3b-d}\right) + ((4-n)b-d)X\right)
\end{aligned}$$

and

$$\begin{aligned}
\Pi^r_n &= -(2b-d)\frac{b(3b-d)}{((2+n)b-d)^2}X\left(\frac{b(1-n)}{(2+n)b-d}X - \frac{a-c}{3b-d}\right) \\
\Pi^b_n &= \frac{b(3b-d)}{((2+n)b-d)^2}X\left(\left(\frac{a-c}{3b-d}\right) + (2b-d)X\right)
\end{aligned}$$

we have

$$\Pi^r_X(0,n) = \Pi^b_X(0,n) = \frac{b(1-n)(2b-d)}{(2+n)b-d}\frac{a-c}{3b-d}$$
$$\text{and } \Pi^r_n(0,n) = \Pi^b_n(0,n) = 0.$$

The Jacobian matrix evaluated at the equilibrium then is

$$\mathbf{J}^* = \begin{pmatrix} -\frac{b(1-n)}{(2+n)b-d} & 0 \\ g_X & (1-2\delta)\frac{\Pi^*(\Pi^*-T)}{(\Pi^*-nT)^2} \end{pmatrix},$$

with eigenvalues

$$\lambda_1 = -\frac{b(1-n)}{(2+n)b-d} \text{ and } \lambda_2 = (1-2\delta)\frac{\Pi^*(\Pi^*-T)}{(\Pi^*-nT)^2}.$$

and corresponding eigenvectors

$$v_1 = \begin{pmatrix} \lambda_1 - \lambda_2 \\ \frac{a-c}{3b-d}\frac{(2b-d)n(1-n)T}{\Pi^*(\Pi^*-T)}\lambda_1\lambda_2 \end{pmatrix} \text{ and } v_2 = \begin{pmatrix} 0 \\ 1 \end{pmatrix}.$$

The product of the eigenvalues has absolute value

$$|\lambda_1 \times \lambda_2| = (1-2\delta)\frac{b(1-n)}{(2+n)b-d}\frac{\Pi^*(\Pi^*-T)}{(\Pi^*-nT)^2}$$

which, for δ and n_δ^* close to 0, is approximately $\frac{b}{2b-d}\frac{\Pi^*-T}{\Pi^*}$.

5.8.4 Some Important Curves

In this section we derive some of the curves that play an important role in the analysis of our model. We start with the main computations necessary to determine L_δ^{III}, the third component of the stable manifold. Using the fact that the profit functions are quadratic in X we can derive the following solutions

$$X_{1,2} = \frac{1}{2a_n}\left(-b_n \pm \sqrt{b_n^2 - 4a_n c_n}\right)$$

where

$$\begin{aligned} a_n &= \mu n\left(\frac{b(1-n)}{(2+n)b-d}\right)^2 - (1-n)\left(\frac{(4-n)b-d}{(2+n)b-d}\right), \\ b_n &= -2((\mu-1)n+1)\left(\frac{b(1-n)}{(2+n)b-d}\right)\left(\frac{a-c}{3b-d}\right) \text{ and} \\ c_n &= ((\mu-1)n+1)\left(\frac{a-c}{3b-d}\right)^2 - \frac{\mu n}{b-\frac{1}{2}d} \end{aligned}$$

and where $\mu = \frac{\delta}{1-\delta}$. These points solve

$$1 = (1-2\delta)\frac{n(\Pi^r(X,n)-1)}{n(\Pi^r(X,n)-1)+(1-n)\Pi^b(X,n)} + \delta.$$

Clearly for $\delta = 0$, there is no solution to this equation. However, it can be easily shown that $\lim_{n\to 0} X(n) = \frac{2b-d}{4b-d}\frac{a-c}{3b-d} \equiv X^f$.

Next we focus on the zero-profit curves L^b and L^r. Since $\Pi^b(X,n)$ and $\Pi^r(X,n)$ are quadratic functions in X the zero-profit functions can be easily solved for X, given n. The relevant solutions are given by

$$\begin{aligned} L^b &= \left\{(X,n)\,|X = \frac{(2+n)\,b-d}{(4-n)\,b-d}\frac{a-c}{3b-d}\right\} \text{ and} \\ L^r &= \left\{(X,n)\,|X = \frac{(2+n)\,b-d}{b\,(1-n)}\left(\frac{a-c}{3b-d}-\sqrt{\frac{2T}{2b-d}}\right)\right\}. \end{aligned}$$

Now consider the curve L^n, the curve consisting of the points at which the denominator of $g(X,n)$ vanishes. This occurs when

$$n\left(\Pi^r(X,n)-\Pi^b(X,n)\right)-nT+\Pi^b(X,n)=0.$$

Again the right-hand side of this equation is quadratic in X and therefore can be solved. Now define

$$y=\frac{1}{(2+n)\,b-d}X \text{ and } z=\frac{a-c}{3b-d}$$

Then the relevant solution is

$$L^n=\{(X,n)\,|X=((2+n)\,b-d)\,y_n\}$$

where

$$\begin{aligned} y_{1,2} &= \frac{1}{(1-n)\left((1-n)\,b^2-(3b-d)^2\right)}\left(zb\,(1-n)-\sqrt{D}\right) \text{ and} \\ D &= (1-n)\left((3b-d)^2 z^2+\left((1-n)\,b^2-(3b-d)^2\right)n\frac{2T}{2b-d}\right). \end{aligned}$$

5.8.5 Diffeomorphisms

From the inverse function theorem it follows that a map has a local smooth inverse if the determinant is unequal to 0. So to determine where a local smooth inverse exists we want to find the set of points (X,n) at which the determinant vanishes. For our dynamical system $F(X,n)=(f(X,n),g(X,n))'$ we have (as computed in appendix 5.8.3) that $\det J(X,n)$ equals the expression

$$-\frac{(1-2\delta)\,b\,(1-n)}{(2+n)\,b-d}\frac{\Pi^b(\Pi^r-T)+n\,(1-n)\left(\Pi^b\Pi^r_n-(\Pi^r-T)\,\Pi^b_n\right)}{\left(n\,(\Pi^r-T)+(1-n)\,\Pi^b\right)^2}$$
$$-\frac{(1-2\delta)\,b\,(3b-d)}{((2+n)\,b-d)^2}X\frac{n\,(1-n)\left[\Pi^b\Pi^r_X-(\Pi^r-T)\,\Pi^b_X\right]}{\left(n\,(\Pi^r-T)+(1-n)\,\Pi^b\right)^2}$$

Since all terms contain the term $(1-n)$, we know that the determinant vanishes at $n=1$. Now assume $n<1$ and $n(\Pi^r(X,n)-T)+(1-n)\Pi^b(X,n)\neq 0$ and divide $\det J(X,n)$ by the strictly positive number $(1-2\delta)(1-n)\times\left(n(\Pi^r-T)+(1-n)\Pi^b\right)^{-2}$. Then the determinant equals zero when the expression $D(X,n)=$

$$-((2+n)b-d)\left(\Pi^b(\Pi^r-T)+n(1-n)\left(\Pi^b\Pi^r_n-(\Pi^r-T)\Pi^b_n\right)\right) \\ -(3b-d)Xn\left[\Pi^b\Pi^r_X-(\Pi^r-T)\Pi^b_X\right]$$

equals zero. It can be easily checked that for all points with $\Pi^r(X,n)=T$ (that is, points in L^r) the determinant vanishes. We also computed another branch of points where the determinant vanishes numerically. This branch (which we call $L^{\det}$) starts in the focal point $\left(X^f,0\right)$.

Chapter 6

Price Adjustment in Monopolistic Competition

6.1 Introduction

There is a long-standing and still unanswered question in the theory of general equilibrium: how does the economy get to an equilibrium? The best-known adjustment process is the tâtonnement process, which was discussed in Chapter 2. There we saw that the tâtonnement process need not converge to an equilibrium price vector and indeed this process allows for all kinds of complicated dynamical behaviour. Furthermore some undesirable features of the tâtonnement process were discussed: the absence of trade at disequilibrium prices and the problem of price-setting. With respect to the latter, it was suggested that a more realistic model of price adjustment would require that economic agents themselves, instead of an abstract institution such as the *auctioneer* would set prices.[1] This issue is explored further in the present chapter. This chapter reviews some simple adjustment processes in partial equilibrium models of monopolistic competition. These models may serve as a first (admittedly small) step in the direction of a more satisfactory theory of price adjustment in general equilibrium models.

We are interested in the following type of model. Each commodity is manufactured by a single firm. These firms have incomplete information in the following sense: they do not know the demand for the commod-

[1] Notice that the same argument holds for Cournot-type models (as the one studied in Chapter 5), where firms compete in quantities and where it is unclear how market clearing prices emerge.

ity they produce and they assume that demand only depends upon the price of that commodity, that is, they do not take into account the interdependence with the prices of other commodities. Every individual firm acts as if it is a monopolist for the commodity it produces. Then a learning dynamics of the following type arises. In each period each firm estimates a demand curve for the commodity. This *perceived* demand curve leads to a (perceived) optimal price which in turn leads to more information on the demand curve, leading to a new optimal price, and so on. Notice that the demand curves are in general misspecified, since they do not incorporate information on prices for other products, that is, they believe the rest of the economy is constant. Furthermore, the functional specification might be wrong, when firms, for example, estimate demand curves that are linear in the own price, whereas the true demand curves might be nonlinear in the own price. We study the convergence properties of these learning processes. Two types of learning models are considered in particular. They differ in the information that is used to estimate the demand curves. These two types of learning models are by no means the only possible models of learning, but they seem to be interesting benchmark cases.

In the short memory model only local information is used and learning proceeds as follows. When a firm sets a price it observes the amount it could have sold at that price and it observes the sensitivity of the demand curve with respect to small variations in that price, that is, it knows the derivative of the demand curve at the price it sets (it might observe this sensitivity through small market experiments for example). On the basis of this information a *linear* demand curve is estimated and a new optimal price is determined. In the next period, information about sales and the slope of the demand curve at the new price is gathered and this information gives a new estimated demand curve, and so on.

The long memory learning process we discuss has been studied by Kirman (1975, 1983) and is based on the following idea. Again firms collect information on price and quantity combinations from their perceived demand curve. In this case they do not know the sensitivity of the demand with respect to the own price, but they use all previous information to estimate a demand function. Firms use a recursive ordinary least squares algorithm to estimate this demand curve. Then there seems to be convergence on a so-called *conjectural* equilibrium. Convergence of these kinds of learning processes in a more general framework has been shown by Schinkel, Tuinstra and Vermeulen (1999).

A next step in the analysis would be to incorporate these learning processes into a general equilibrium framework and eventually allow for trade at disequilibrium prices. This is left for future research. We want to mention one important point here briefly: is it reasonable to assume that agents act as monopolists? This appears to be a less severe restriction in a general equilibrium model where, for example, each sector can be represented by a single firm that acts monopolistically (see Nikaido (1975)) than in a partial equilibrium model where this monopolistic behaviour is on the level of differentiated commodities, where there is more mutual dependency between the demand curves for the different commodities.

The outline for the rest of this chapter is as follows. In Section 6.2 we discuss the partial equilibrium model and some well-known adjustment processes are reviewed. Section 6.3 is devoted to the adjustment process in a partial equilibrium model, where firms know the derivative of the demand functions for any price they quote. Section 6.4 deals with the long-memory learning process of Kirman. Section 6.5 summarizes. Most of the results discussed in this chapter, except for the short run learning process in Section 6.3, are known. This chapter should therefore mainly be seen as an overview of the literature on price adjustment processes in models of monopolistic competition.

6.2 A partial equilibrium model

We consider a partial equilibrium model with heterogeneous commodities and imperfect competition. Let there be n firms, where each firm produces its own unique commodity and is the sole supplier of that commodity. The costs of making this product for firm i are given by a nonnegative, continuous, nondecreasing and twice differentiable function $C_i : I\!R_+ \to I\!R_+$. Let $\mathbf{p} = (p_1, \ldots, p_n) \in I\!R^n_+$ be a price vector, where p_i is the price for the i'th commodity. Occasionally we write the price vector as $(p_i, \mathbf{p}_{-i})$, where $\mathbf{p}_{-i} = (p_1, \ldots, p_{i-1}, p_{i+1} \ldots, p_n)$. The demand for commodity i depends upon the prices of all commodities and is given by a demand function $D^i : I\!R^n_+ \to I\!R_+$, which is assumed to be nonnegative, continuous and twice differentiable in all its arguments, whenever it is strictly positive. First and second order derivatives of the demand functions are denoted by $D^i_j(\mathbf{p}) \equiv \frac{\partial D^i(\mathbf{p})}{\partial p_j}$ and $D^i_{jk}(\mathbf{p}) \equiv \frac{\partial^2 D^i(\mathbf{p})}{\partial p_j \partial p_k}$. We assume demand is nonincreasing in the own price ($D^i_i(\mathbf{p}) \leq 0$) and that $D^i_j(\mathbf{p})$ and $D^j_i(\mathbf{p})$ have the same sign. If $D^i_j(\mathbf{p})$ and $D^j_i(\mathbf{p})$ are positive we call commodities i and j *substitutes*, if they are negative, we call them *complements*. Each firm chooses its own price in order to maximize profits, which are given by

$$\pi_i(p_i, \mathbf{p}_{-i}) = p_i D^i(\mathbf{p}) - C_i\left(D^i(\mathbf{p})\right). \tag{6.1}$$

We can then define a *Bertrand-Nash* equilibrium as follows.

Definition 6.1 *A price vector* $\mathbf{p}^* = (p_1^*, \dots, p_n^*)$ *is a* Bertrand-Nash equilibrium *if for each* $i = 1, \dots, n$ *we have*

$$\pi_i\left(p_i^*, \mathbf{p}_{-i}^*\right) \geq \pi_i\left(p_i, \mathbf{p}_{-i}^*\right) \text{ for all } p_i.$$

The idea underlying this equilibrium concept is that, given that all firms charge the Bertrand-Nash price, no firm can increase its profit unilaterally by deviating from this price level. Average payoff can be increased though, if firms cooperate and therefore the equilibrium will in general not be Pareto optimal with respect to firms. An alternative formulation of the Bertrand-Nash equilibrium is in terms of the *reaction functions.* The reaction function for firm i gives the price that maximizes his profit, given the prices set by the other firms, that is, this reaction curve is given by

$$R^i(\mathbf{p}_{-i}) = \arg\max_{p_i} \pi(p_i, \mathbf{p}_{-i}).$$

A Bertrand-Nash equilibrium then corresponds to a fixed point of the map

$$\mathbf{R}(\mathbf{p}) = \begin{pmatrix} R^1(\mathbf{p}_{-1}) \\ \vdots \\ R^n(\mathbf{p}_{-n}) \end{pmatrix}.$$

To guarantee the existence of a Bertrand-Nash equilibrium we have to assume quasiconcavity of the profit functions, a condition which cannot be derived from assumptions on the fundamentals of the economy, such as preferences, endowments or technology. In particular, as can be seen by differentiating (6.1) twice, this requires the demand function not to be too convex. However, it is well known (see e.g. Sonnenschein (1973)) that demand functions can take almost any form and still be consistent with utility maximization. As was pointed out in an influential paper by Roberts and Sonnenschein (1977), lack of quasiconcavity of the profit functions may cause the reaction curves to exhibit discontinuities, possibly leading to nonexistence of the Bertrand-Nash equilibrium (nonpathological and robust examples can be found in e.g. Roberts and Sonnenschein (1977), Friedman (1983), Novshek (1985) and Bonanno (1988)).

A way out of this problem would be to consider equilibria along the lines of Bonanno and Zeeman (1985) and Bonanno (1988). They consider price setting oligopolies where all agents have constant marginal costs, that is $C_i(x_i) = c_i x_i$. Bonanno and Zeeman (1985) consider the case where producers only focus on the first order conditions for an optimum, that is, a price vector $\mathbf{p}^* = (p_1^*, \ldots, p_n^*)$ is called an *equilibrium* when

$$\frac{\partial \pi_i(\mathbf{p}^*)}{\partial p_i} = 0, \quad \text{for all } i. \tag{6.2}$$

Bonanno and Zeeman show that under some very mild conditions, such an equilibrium always exists. At such an equilibrium it is however possible that some producer is at a local minimum of his profit function. It is somewhat peculiar to call such a situation an equilibrium since all firms disregard second order conditions, but it can be defended on the grounds that if the producers think they face a linear demand curve, they believe they are at the global maximum of their profit function whenever the first order condition is satisfied. Bonanno (1988) defines a *local Bertrand-Nash* equilibrium as a price vector $\mathbf{p}^* = (p_1^*, \ldots, p_n^*)$ for which, besides (6.2) the second order condition for a local maximum is satisfied, that is,

$$\frac{\partial^2 \pi_i(\mathbf{p}^*)}{\partial p_i^2} < 0, \quad \text{for all } i. \tag{6.3}$$

At such a local Bertrand-Nash equilibrium no producer can improve his profit by deviating to a price in the *neighbourhood* of the equilibrium price. A sufficient condition for such a local Bertrand-Nash equilibrium to exist is that $D_{ii}^i(\mathbf{p}) \neq 2D^i(\mathbf{p}) \big/ (p_i - c_i)^2$ for all i and $\mathbf{p}$ at which demand is positive. This condition implies that the first and second order derivative of the profit function never vanish together, which excludes the case that a local minimum and a local maximum of the profit function merge.

6.2.1 Best-reply dynamics

We are interested in how the firms might coordinate on a Bertrand-Nash equilibrium. There is a large literature analyzing the stability of the best-reply dynamics in the homogeneous Cournot oligopoly model. In the homogeneous Cournot oligopoly model there is just one commodity and therefore one price. This price is the market clearing price and is determined by the total output supplied by the firms. Notice that for the Cournot model, just as for the general equilibrium model, it is unclear how this market clearing price emerges. The underlying idea of the best-reply dynamics is the following. Each individual firm assumes the outputs of the other firms to remain unchanged. It then produces the amount which, given these fixed quantities of its competitors, maximizes its profit. That is, it produces the amount specified by its reaction curve. Each firm therefore has full information with respect to its environment, that is, it knows the inverse demand function and previous actions of its competitors, but it has to predict the amount that its competitors will produce in the current period. The study of the stability of the homogeneous Cournot oligopoly model in modern times[2] started with Theocharis (1959), who showed that given a linear inverse demand function and constant marginal costs, the Cournot-Nash equilibrium is stable if there are only 2 firms, but that bounded oscillations occur when $n = 3$ and that for $n > 3$ quantities diverge.[3] This result provoked a number of other papers on the stability of the best-reply dynamics for the homogenous Cournot oligopoly model. Here we briefly discuss the well-known papers by Fisher (1961) and Hahn (1962).

Fisher (1961) extends the model of Theocharis in two directions: he considers a quadratic cost function which implies that marginal costs might be decreasing and he introduces a gradual adjustment to the best-reply quantity. He then shows that speeds of adjustment matter in this model. Furthermore, considering the continuous (limit) version of the model he shows that, provided that the slope of the linear inverse demand function is lower than the slope of the marginal cost function (which is positive if the cost function is convex), this continuous adjustment process is stable. Then, for speeds of adjustment sufficiently low, the discrete adjustment process is also stable. It remains however true that an increase in the number of firms might destabilize the best-reply dynamics. Similar results were obtained by McManus and Quandt

[2] These best-reply dynamics were for the first time mentioned by Cournot himself (Cournot (1838)).

[3] Recall that we studied a Cournot model with $n = 2$ and a linear inverse demand function in Chapter 5. The Cournot-Nash equilibrium was ustable under the best-reply dynamics. This instability was due to the fact that marginal costs were decreasing for that example.

(1961).

Hahn (1962) considers the global stability of the continuous adjustment process by using the theory on Lyapunov functions. Hahn considers a general specification of the Cournot oligopoly model and shows that the continuous best-reply dynamics are globally stable under the following two conditions: *i*) the slope of the inverse demand function has to be smaller than the slope of the marginal cost function for any producer (which is the condition derived by Fisher (1961)) and *ii*) marginal revenue of each firm should fall, if total output (given output of this firm) increases.[4] However, al-Nowaihi and Levine (1985) show that there is an error in Hahn's proof and claim that his results are only valid when the number of firms is less than 6. Seade (1980) argues that Hahn's second condition is rather severe and shows that, if it does not hold, instability is a general feature of the continuous adjustment process.

The literature discussed above deals with the homogeneous Cournot model. Here we will briefly describe analogous results for the best-reply dynamics in the heterogeneous Bertrand oligopoly model set out above. We assume reaction curves are continuous and differentiable and a (not necessarily unique) Bertrand-Nash equilibrium $\mathbf{p}^*$ exists. The *instantaneous* best-reply dynamics are given by

$$\begin{pmatrix} p_{1,t+1} \\ \vdots \\ p_{n,t+1} \end{pmatrix} = \begin{pmatrix} R^1(\mathbf{p}_{-1,t}) \\ \vdots \\ R^n(\mathbf{p}_{-n,t}) \end{pmatrix}. \tag{6.4}$$

These best-reply dynamics constitute an $n-$dimensional dynamical system.

Proposition 6.1 *The Bertrand-Nash equilibrium* $\mathbf{p}^*$ *is locally stable under the instanteneous best-reply dynamics (6.4) if*

$$\left|\frac{\partial^2 \pi_i(\mathbf{p}^*)}{\partial p_i^2}\right| > \sum_{j \neq i} \left|\frac{\partial^2 \pi_i(\mathbf{p}^*)}{\partial p_i \partial p_j}\right|, \quad \textit{for all } i. \tag{6.5}$$

Proof. The equilibrium $\mathbf{p}^*$ is locally stable if the eigenvalues of the Jacobian matrix of (6.4) evaluated at $\mathbf{p}^*$ lie in the unit circle. This

---

[4]That is, if $p(X)$ is the inverse demand function where $X = x_1 + \ldots + x_n$, then we must have $p'(X) + x_i p''(X) < 0$, for all i and all X.

Jacobian matrix is $\mathbf{J} =$

$$\begin{pmatrix} 0 & R_2^1\left(\mathbf{p}_{-1}^*\right) & R_3^1\left(\mathbf{p}_{-1}^*\right) & \cdots & & R_n^1\left(\mathbf{p}_{-1}^*\right) \\ R_1^2\left(\mathbf{p}_{-2}^*\right) & 0 & & & & R_n^2\left(\mathbf{p}_{-2}^*\right) \\ \vdots & & \ddots & & & \\ \vdots & & & & 0 & R_n^{n-1}\left(\mathbf{p}_{-(n-1)}^*\right) \\ R_1^n\left(\mathbf{p}_{-n}^*\right) & R_1^n\left(\mathbf{p}_{-n}^*\right) & & & R_{n-1}^n\left(\mathbf{p}_{-n}^*\right) & 0 \end{pmatrix},$$

where $R_j^i\left(\mathbf{p}_{-i}\right) \equiv \frac{\partial R^i(\mathbf{p}_{-i})}{\partial p_j}$. Now we use the following result on matrices to derive a sufficient condition for local stability (see e.g. Atkinson (1989)). Let $\|\cdot\|$ be a matrix norm and let $\|\cdot\|_v$ be a vector norm *compatible* with $\|\cdot\|$, i.e. $\|\mathbf{Ax}\|_v \leq \|\mathbf{A}\| \, \|\mathbf{x}\|_v$. Let λ be an eigenvalue of the matrix $\mathbf{A}$ with corresponding eigenvector $\mathbf{x}$. By the definition of a norm we have $\|\mathbf{Ax}\|_v = \|\lambda \mathbf{x}\|_v = |\lambda| \, \|\mathbf{x}\|_v$. Combining these two gives $|\lambda| \, \|\mathbf{x}\|_v \leq \|\mathbf{A}\| \, \|\mathbf{x}\|_v \Rightarrow |\lambda| \leq \|\mathbf{A}\|$. So the largest eigenvalue of $\mathbf{A}$ is always smaller, in absolute value, than the norm of $\mathbf{A}$. Now consider the maximum row sum norm $\|\mathbf{A}\|_\infty = \max_i \sum_{j=1}^n |\mathbf{A}_{ij}|$, i.e. the largest row sum of absolute values of elements of $\mathbf{A}$. The vector norm $\|\mathbf{x}\|_\infty = \max_i |\mathbf{x}_i|$ is compatible with this matrix norm. A sufficient condition for the Bertrand-Nash equilibrium to be locally stable under the best-reply dynamics then becomes $\sum_{j=1}^n \left| R_j^i\left(\mathbf{p}_{-i}^*\right)\right| < 1$ for all i. $R^i\left(\mathbf{p}_{-i}\right)$ is implicitly defined by $\frac{\partial \pi_i\left(R^i(\mathbf{p}_{-i}),\mathbf{p}_{-i}\right)}{\partial p_i} = 0$. Totally differentiating this equation with respect to p_j and solving for $R_j^i\left(\mathbf{p}_{-i}\right) \equiv \frac{\partial R^i(\mathbf{p}_{-i})}{\partial p_j}$ gives

$$R_j^i\left(\mathbf{p}_{-i}\right) = -\frac{\partial^2 \pi_i\left(\mathbf{p}\right)}{\partial p_i \partial p_j} \Bigg/ \frac{\partial^2 \pi_i\left(\mathbf{p}\right)}{\partial p_i^2}. \tag{6.6}$$

The condition $\sum_{j=1}^n \left| R_j^i\left(\mathbf{p}_{-i}^*\right)\right| < 1$ then becomes equivalent with (6.5) ■

For the case of linear cost functions, that is $C_i\left(x_i\right) = c_i x_i$, the first order condition for an optimum is

$$\left(p_i^* - c_i\right) D_i^i\left(\mathbf{p}^*\right) + D^i\left(\mathbf{p}^*\right) = 0, \qquad \text{for all } i. \tag{6.7}$$

Condition (6.5) then becomes

$$\left|2D_i^i\left(\mathbf{p}^*\right) + \left(p_i^* - c_i\right) D_{ii}^i\left(\mathbf{p}^*\right)\right| > \sum_{j \neq i} \left|D_j^i\left(\mathbf{p}^*\right) + \left(p_i^* - c_i\right) D_{ij}^i\left(\mathbf{p}^*\right)\right|.$$

Using (6.7) this reduces to

$$\left|\frac{2\left[D_i^i(\mathbf{p}^*)\right]^2 - D^i(\mathbf{p}^*)\,D_{ii}^i(\mathbf{p}^*)}{D_i^i(\mathbf{p}^*)}\right| >$$

$$\sum_{j\neq i}\left|\frac{D_j^i(\mathbf{p}^*)\,D_i^i(\mathbf{p}^*) - D^i(\mathbf{p}^*)\,D_{ij}^i(\mathbf{p}^*)}{D_i^i(\mathbf{p}^*)}\right|.$$

If condition (6.5) holds not only at the equilibrium, but at any price vector $\mathbf{p}$, the equilibrium $\mathbf{p}^*$ is unique and globally stable (Friedman (1977)). It suffices to show that the mapping $\mathbf{R}(\mathbf{p})$ is a contraction, that is, $\|\mathbf{R}(\mathbf{p}) - \mathbf{R}(\mathbf{q})\| < \|\mathbf{p} - \mathbf{q}\|$ for all $\mathbf{p}$ and $\mathbf{q}$, since it is well-known that a contraction has a unique and globally stable fixed point. The desired result is then given by the fact that $\mathbf{R}(\mathbf{p})$ is a contraction if $\sum_{j=1}^n \left|R_j^i(\mathbf{p}_{-i})\right| < 1$ for all i and all $\mathbf{p}$. Notice that this condition is far more demanding than the condition for local stability, since it has to be satisfied at every price vector.

For the sake of completeness we briefly turn to the study of the continuous best-reply dynamics, which are given by

$$\frac{dp_i}{dt} = K_i\left(R^i(\mathbf{p}_{-i}) - p_i\right), \qquad i = 1, \ldots, n. \tag{6.8}$$

Here $K_i > 0$ is the adjustment speed of producer i.

Proposition 6.2 *The Bertrand-Nash equilibrium* $\mathbf{p}^*$ *is globally stable under the continuous best-reply dynamics (6.8) if*

$$\left|\frac{\partial^2 \pi_i(\mathbf{p})}{\partial p_i^2}\right| > \sum_{j\neq i}\left|\frac{\partial^2 \pi_i(\mathbf{p})}{\partial p_i \partial p_j}\right|, \text{ for all } \mathbf{p}, \text{ and all } i. \tag{6.9}$$

Proof. Consider the following Lyapunov function

$$V(\mathbf{p}) = \max_i\left\{K_i\left|R^i(\mathbf{p}) - p_i\right|\right\}.$$

Clearly, $V(\mathbf{p}) = 0$ if and only if $\mathbf{p}$ is a Bertrand-Nash equilibrium and $V(\mathbf{p}) > 0$ otherwise. If we can show that $\frac{dV}{dt} < 0$ for all $\mathbf{p} \neq \mathbf{p}^*$, we know that the continuous best-reply dynamics are globally stable. Now let i be such that for all j we have

$$K_i\left|R^i(\mathbf{p}) - p_i\right| \geq K_j\left|R^j(\mathbf{p}) - p_j\right|.$$

First assume $R^i - p_i > 0$. Then we have

$$\begin{aligned}\frac{dV}{dt} &= K_i\left(\frac{dR^i}{dt} - \frac{dp_i}{dt}\right) = -K_i^2\left(R^i - p_i\right) + K_i\frac{dR^i}{dt} \\ &= -K_i^2\left(R^i - p_i\right) - \sum_{j\neq i} q_{ij}K_iK_j\left(R^j - p_j\right),\end{aligned}$$

where we have used

$$\frac{dR^i}{dt} = -\sum_{j\neq i} q_{ij}\frac{dp_j}{dt}, \text{ where } q_{ij} = \frac{\partial^2\pi_i(\mathbf{p})}{\partial p_i \partial p_j} \Big/ \frac{\partial^2 \pi_i(\mathbf{p})}{\partial p_i^2}.$$

If $\sum_{j\neq i}|q_{ij}| < 1$ we obtain

$$\sum_{j\neq i} q_{ij}K_iK_j\left(R^j - p_j\right) \leq \sum_{j\neq i}|q_{ij}|\, K_i^2\left(R^i - p_i\right) < K_i\left(R^i - p_i\right)^2,$$

where the first inequality follows from the assumption that $K_i\left(R^i - p_i\right) \geq \left|K_j\left(R^j - p_j\right)\right|$ for all j. Now consider $R^i - p_i < 0$. Then

$$\frac{dV}{dt} = -K_i\left(\frac{dR^i}{dt} - \frac{dp_i}{dt}\right) = K_i^2\left(R^i - p_i\right) + \sum_{j\neq i} q_{ij}K_iK_j\left(R^j - p_j\right).$$

Then if $\sum_{j\neq i}|q_{ij}| < 1$ and using that $-K_i\left(R^i - p_i\right) \geq \left|K_j\left(R^j - p_j\right)\right|$ for all j we obtain

$$\sum_{j\neq i} q_{ij}K_iK_j\left(R^j - p_j\right) \leq -\sum_{j\neq i}|q_{ij}|\, K_i^2\left(R^i - p_i\right) < -K_i^2\left(R^i - p_i\right).$$

Therefore $\sum_{j\neq i}|q_{ij}| < 1$ guarantees global stability of the (unique) Bertrand-Nash equilibrium. This condition is equivalent with (6.9). ■

This condition is exactly the same as the one we found for the instantaneous best-reply dynamics. Notice that all conditions derived in this section are *sufficient* and not *necessary* conditions. Hence these conditions might be too demanding.

It has been known for some time that the best-reply dynamics can also generate periodic and chaotic patterns. The crucial reference here is Rand (1978), who shows that periodic and chaotic behaviour can occur in the homogenous Cournot duopoly model. Other papers showing this kind of behaviour are Dana and Montrucchio (1986), Bischi and Gardini

(1997) and Kopel (1997). As an example consider the situation with $n = 2$. A crucial aspect of the reaction curves is that they are unimodal, that is, there exists a critical value p_2^c of p_2, such that for $p_2 < p_2^c$, an optimal reaction to an increase in p_2 is an increase in p_1 and for $p_2 > p_2^c$ an optimal reaction to an increase in p_2 is a decrease in p_1. So, for small prices the products are complements and for high prices they are substitutes. This situation might be conceivable when demand is formed in the following way. Consumers are inclined to spend only a certain amount on the type of product supplied by these firms. Furthermore, they prefer having both varieties of the product over having a lot of one variety. Therefore, for low prices consumers want to spend money on both varieties and commodities are complements, whereas for high prices individual consumers have to choose between the varieties and on an aggregate level the varieties are substitutes. If the reaction curves are indeed unimodal, periodic and chaotic behaviour might occur.[5]

6.2.2 Gradient systems

Another type of adjustment processes studied in the Cournot model are gradient processes, see for example Arrow and Hurwicz (1960), Furth (1986), Bischi and Naimzada (1995) and Corchón and MasCollel (1997). These gradient processes assume that firms change their price in the direction in which profit increases. That is, if marginal profit at the current price is positive, the firm increases its price and if marginal profit is negative, the firm decreases its price. In both cases the firm expects an increase in profits, however, since other firms also change their prices, it is very well possible that profits in fact decrease. In continuous time this process can be written as

$$\frac{dp_i}{dt} = K_i \frac{\partial \pi_i}{\partial p_i}, \qquad i = 1, \ldots, n, \tag{6.10}$$

where, as before, K_i is the speed with which the i'th producer changes the price for its product. First consider the steady states of this gradient dynamics. Any price vector where the first order conditions for a maximum are satisfied for each producer is a steady state of this dynamics.

[5] It is very easy to generate such behaviour. Assume zero costs and the following demand function for commodity i : $D^i(p_i, p_j) = f_i(p_j) - \frac{1}{2}p_i$, $j \neq i$. Then the reaction curve for firm i is given by $p_i = f_i(p_j)$. Then for example, if we let $f_i(p_j) = \lambda_i p_j (1 - p_j)$, the best-reply dynamics is equivalent to the double logistic map, which exhibits all kinds of complicated periodic and chaotic behaviour.

This implies that next to Bertrand-Nash equilibria and local Nash equilibria we might also have steady states where some producer is at a local minimum of his profit function. For any such a price vector we have the following result. The Jacobian matrix is

$$J = \begin{pmatrix} K_1 \frac{\partial^2 \pi_1}{\partial p_1^2} & K_1 \frac{\partial^2 \pi_1}{\partial p_1 \partial p_2} & \cdots & K_1 \frac{\partial^2 \pi_1}{\partial p_1 \partial p_n} \\ K_2 \frac{\partial^2 \pi_2}{\partial p_2 \partial p_1} & K_2 \frac{\partial^2 \pi_2}{\partial p_2^2} & & \\ \vdots & & \ddots & \\ K_n \frac{\partial^2 \pi_n}{\partial p_n \partial p_1} & & & K_n \frac{\partial^2 \pi_n}{\partial p_n^2} \end{pmatrix}.$$

If this matrix has diagonal dominance, that is, if for each row, the diagonal elements are larger in absolute value than the sum of the absolute values of the off-diagonal elements, and if these diagonal elements are negative, then J is negative definite. This implies that the real parts of the eigenvalues are negative and hence the gradient process is locally stable. The condition for local stability for a (local) Bertrand-Nash equilibrium therefore becomes

$$\left| \frac{\partial^2 \pi_i (\mathbf{p}^*)}{\partial p_i^2} \right| > \sum_{j \neq i} \left| \frac{\partial^2 \pi_i (\mathbf{p}^*)}{\partial p_i \partial p_j} \right|, \qquad \text{for all } i,$$

which is condition (6.5) from the previous section. For a steady state corresponding to a local minimum of the profit function, the situation is somewhat more complex. We then in general expect the system to be locally unstable.

Now consider global stability.

Proposition 6.3 *The Bertrand-Nash equilibrium* $\mathbf{p}^*$ *is globally stable under the gradient system (6.10) if*

$$-\frac{\partial^2 \pi_i (\mathbf{p})}{\partial p_i^2} > \sum_{j \neq i} \left| \frac{\partial^2 \pi_i (\mathbf{p})}{\partial p_i \partial p_j} \right|, \text{ for all } i \text{ and all } \mathbf{p}. \tag{6.11}$$

Proof. Take the Lyapunov function

$$V(\mathbf{p}) = \max_i \left\{ K_i \left| \frac{\partial \pi_i}{\partial p_i} \right| \right\}$$

and let i be such that

$$K_i \left| \frac{\partial \pi_i}{\partial p_i} \right| \geq K_j \left| \frac{\partial \pi_j}{\partial p_j} \right|, \text{ for all } j.$$

If $K_i \frac{\partial \pi_i}{\partial p_i} > 0$ we have

$$\begin{aligned}\frac{dV}{dt} &= K_i \sum_{j=1}^{n} \frac{\partial^2 \pi_i}{\partial p_i \partial p_j} \frac{dp_j}{dt} = K_i \sum_{j=1}^{n} \frac{\partial^2 \pi_i}{\partial p_i \partial p_j} K_j \frac{\partial \pi_j}{\partial p_j} \\ &\leq \frac{\partial^2 \pi_i}{\partial p_i^2} K_i^2 \frac{\partial \pi_i}{\partial p_i} + \sum_{j \neq i} \left| \frac{\partial^2 \pi_i}{\partial p_i \partial p_j} \right| K_i^2 \frac{\partial \pi_i}{\partial p_i}\end{aligned}$$

and if $K_i \frac{\partial \pi_i}{\partial p_i} < 0$ we in a similar fashion find

$$\frac{dV}{dt} \leq -\frac{\partial^2 \pi_i}{\partial p_i^2} K_i^2 \frac{\partial \pi_i}{\partial p_i} - \sum_{j \neq i} \left| \frac{\partial^2 \pi_i}{\partial p_i \partial p_j} \right| K_i^2 \frac{\partial \pi_i}{\partial p_i}.$$

Hence a sufficient condition for global stability is (6.11). ■

Notice that this is the same condition as was found for the best-reply dynamics.

As with the best-reply dynamics gradient systems can exhibit complicated behaviour. In fact Corchón and MasCollel (1995) show that any set of functions can be generated by the gradient system with demand functions that satisfy nice properties such as gross substitutability, that is, for which, for example the centralized tâtonnement process is globally stable.

6.3 A short memory learning procedure

In this section we introduce an alternative adjustment process for the partial equilibrium model with imperfect competition set out in the previous section. We assume firms do not know the demand curves for the products they are manufacturing but they perceive these demand curves to only depend upon their own price. This implies that they abstract from any interdependence with prices set by other firms. Furthermore, they believe that this relationship between price and demand is linear. Each firm estimates his demand curve on the basis of *local* information about the demand curve obtained in the previous period. This local information consists of two parts: a firm knows, given the price he charged in the previous period how much he could have sold against that price, and he knows the sensitivity of the demand curve with respect to price changes at that price (that is the slope of the demand curve at that price), for example through some level of experimentation or market research (compare Silvestre (1977)). This information is enough to identify a linear demand curve. On the basis of this perceived linear demand curve an optimal price is determined. In the next period this new price leads to new information about the location and slope of the true demand curve and hence to a new estimate of the perceived demand curve, leading to a new optimal price again. The main question now is if this learning process converges to some sort of equilibrium price vector.

We assume marginal costs are linear, that is $C_i(x_i) = c_i x_i$. Firm i's perceived demand curve is

$$d_i(p_i) = a_i - b_i p_i, \quad \text{with} \quad a_i, b_i > 0. \tag{6.12}$$

Given this perceived demand curve and the cost structure, firm i's (perceived) profit maximizing price is

$$p_i = \frac{a_i}{2b_i} + \frac{1}{2}c_i. \tag{6.13}$$

In each period firm i estimates the demand curve on the basis of information from the last period. Firm i knows the price it charged in the last period and the amount it (could have) sold, so it knows $\left(p_{it}, D^i(\mathbf{p}_t)\right)$. Furthermore it knows the slope of the demand curve in this point, $D^i_i(\mathbf{p}_t)$. The estimated perceived demand curve in period t then becomes

$$d_{it}(p_i) = a_{it} - b_{it}p_i = \left(D^i(\mathbf{p}_t) - D^i_i(\mathbf{p}_t)\,p_{it}\right) + D^i_i(\mathbf{p}_t)\,p_i. \tag{6.14}$$

The price for the next period then is

$$p_{i,t+1} = F_i(\mathbf{p}_t) = \frac{a_{it}}{2b_{it}} + \frac{1}{2}c_i = \frac{1}{2}(p_{it} + c_i) - \frac{1}{2}\frac{D^i(\mathbf{p}_t)}{D^i_i(\mathbf{p}_t)}. \tag{6.15}$$

It can be easily checked that any fixed point of (6.15) satisfies the first order conditions for a Bertrand-Nash equilibrium.

We are interested in the stability of these Bertrand-Nash equilibria. We have the following (recall that $D^i_{jk}(\mathbf{p}) \equiv \frac{\partial^2 D^i(\mathbf{p})}{\partial p_j \partial p_k}$)

Proposition 6.4 *A sufficient condition for the local stability of the Bertrand-Nash equilibrium* $\mathbf{p}^* = (p_1^*, p_2^*, \cdots, p_n^*)$ *is*

$$\left|\frac{D^i(\mathbf{p}^*)\,D^i_{ii}(\mathbf{p}^*)}{\left(D^i_i(\mathbf{p}^*)\right)^2}\right| + \sum_{j \neq i}\left|\frac{D^i(\mathbf{p}^*)\,D^i_{ij}(\mathbf{p}^*) - D^i_i(\mathbf{p}^*)\,D^i_j(\mathbf{p}^*)}{\left(D^i_i(\mathbf{p}^*)\right)^2}\right| \leq 2, \tag{6.16}$$

for all $i = 1, \ldots, n$.

Proof. To determine local stability we have to look at the Jacobian matrix $\mathbf{J}$. The diagonal elements of the Jacobian are

$$J_{ii} = \frac{\partial F_i(\mathbf{p}^*)}{\partial p_i} = \frac{1}{2}\frac{D^i(\mathbf{p}^*)\,D^i_{ii}(\mathbf{p}^*)}{\left(D^i_i(\mathbf{p}^*)\right)^2}$$

The off-diagonal elements are

$$J_{ij} = \frac{\partial F_i(\mathbf{p}^*)}{\partial p_i} = \frac{1}{2}\frac{D^i(\mathbf{p}^*)\,D^i_{ij}(\mathbf{p}^*) - D^i_i(\mathbf{p}^*)\,D^i_j(\mathbf{p}^*)}{\left(D^i_i(\mathbf{p}^*)\right)^2}$$

Let λ be an eigenvalue of $\mathbf{J}$, then, as before we have $|\lambda| \leq \|\mathbf{J}\|$, so the largest eigenvalue in absolute value is smaller than the norm of the Jacobian. If we now take as norm $\|A\|_\infty = \max_i \sum_{j=1}^n |A_{ij}|$, then a sufficient condition for all eigenvalues of the Jacobian matrix to lie in the unit circle is

$$J_i = \sum_{j=1}^{n} |J_{ij}| \leq 1, \text{ for all } i = 1, \ldots, n,$$

and this is equal to condition (6.16). ■

To get some intuition for condition (6.16) notice that the following two conditions are sufficient (but not necessary) for condition (6.16)

$$\left|D^i_i(\mathbf{p}^*)\right| \geq \sum_{j\neq i}\left|D^i_j(\mathbf{p}^*)\right| \text{ and } \left|D^i_i(\mathbf{p}^*)\right| \geq \sqrt{D^i(\mathbf{p}^*)\sum_{j=1}^{n}\left|D^i_{ij}(\mathbf{p}^*)\right|}.$$

The first of these says that the matrix of substitution effects has a dominant diagonal, the second condition says that the curvature of the demand functions has to be small, as compared to the slope. As a last remark notice that condition (6.16) also implies that $\mathbf{p}^*$ satisfies the second order conditions for a Bertrand-Nash equilibrium. That this is so can be seen by rewriting the first and second order conditions for a maximum as

$$\frac{D^i(\mathbf{p})\,D^i_{ii}(\mathbf{p})}{\left(D^i_i(\mathbf{p})\right)^2} \leq 2.$$

We now turn to some examples. Subsection 6.3.1 studies some examples for which convergence takes place and subsection 6.3.2 studies an example for which the learning process might exhibit nonconvergence and for which periodic and strange behaviour might occur.

6.3.1 Examples: convergence

Linear demand functions

Consider a linear system of demand functions of the following form

$$\mathbf{D}(\mathbf{p}) = \boldsymbol{\alpha} + \boldsymbol{\beta}\mathbf{p}, \tag{6.17}$$

where $\boldsymbol{\alpha}$ is a vector of intercepts and $\boldsymbol{\beta}$ is an $n \times n$ matrix, where β_{ii} gives the sensitivity of demand for commodity with respect to its own price and β_{ij} gives the sensitivity of demand for commodity i with respect to the price of commodity j. Assume $\beta_{ii} < 0$.

First consider the Bertrand-Nash equilibrium. The first order conditions for an equilibrium are

$$-2\beta_{ii}p_i = \alpha_i - \beta_{ii}c_i + \sum_{j \neq i} \beta_{ij}p_j, \qquad i = 1, \ldots, n, \tag{6.18}$$

which in matrix notation becomes

$$-\begin{pmatrix} 2\beta_{11} & \beta_{12} & \cdots & \beta_{1n} \\ \beta_{21} & 2\beta_{22} & \cdots & \beta_{2n} \\ \vdots & \vdots & \ddots & \vdots \\ \beta_{n1} & \beta_{n2} & \cdots & 2\beta_{nn} \end{pmatrix} \begin{pmatrix} p_1^* \\ p_2^* \\ \vdots \\ p_n^* \end{pmatrix} = \begin{pmatrix} \alpha_1 - \beta_{11}c_1 \\ \alpha_2 - \beta_{22}c_2 \\ \vdots \\ \alpha_n - \beta_{nn}c_n \end{pmatrix} \tag{6.19}$$

and therefore a nonnegative equilibrium price vector $\mathbf{p}^*$ exists for all $\mathbf{c} = (c_1, \ldots, c_n)' \geq \mathbf{0}$ and all $\boldsymbol{\alpha} \geq \mathbf{0}$ if the matrix $(\boldsymbol{\beta} + \boldsymbol{\beta}_D)^{-1}$ exists and is nonpositive, where $\boldsymbol{\beta}_D$ is a diagonal matrix with β_{ii} as its i'th diagonal element. In particular we have the following

Proposition 6.5 *If $\beta_{ii} < 0$ and $\beta_{ij} \geq 0$ for all i, j with $i \neq j$, then an equilibrium exists for all $\boldsymbol{\alpha} \geq \mathbf{0}$ and all $\mathbf{c} \geq \mathbf{0}$ if*

$$2\left|\beta_{ii}\right| > \sum_{j \neq i} \left|\beta_{ij}\right|, \quad i = 1, \ldots, n. \tag{6.20}$$

Proof. This is a straightforward application of Theorem 4.*C*.3 from Takayama (1985). ■

We now assume that a unique equilibrium exists and is given by

$$\mathbf{p}^* = -(\boldsymbol{\beta} + \boldsymbol{\beta}_D)^{-1}(\boldsymbol{\alpha} - \boldsymbol{\beta}_D\mathbf{c}). \tag{6.21}$$

The adjustment process (6.15) becomes

$$p_{i,t+1} = -\frac{\alpha_i - \beta_{ii}c_i + \sum_{j\neq i}\beta_{ij}p_{j,t}}{2\beta_{ii}}, \qquad i = 1,\ldots,n. \tag{6.22}$$

Notice that (6.22) exactly corresponds to the best-reply dynamics for this case of linear demand functions as can be seen from (6.18). Also, the linear specification (6.17) together with the linear perceived demand curves implies that each firm knows the slope of the demand curve (β_{ii} for firm i) but does not know the intercept since it neglect cross-price effects. The following stability property can easily be checked by looking at (6.22).

Corollary 6.1 *If a unique equilibrium* $\mathbf{p}^* \geq 0$ *exists and if the* β_{ij}*'s satisfy condition (6.20) the dynamical system is globally stable.*

Proof. Condition (6.20) follows directly from condition (6.16). Since the dynamical system is linear, local stability implies global stability. ■

Notice that condition (6.20) is weaker than diagonal dominance.

Loglinear demand functions and linear cost functions

As another special case consider the situation where firms face demand functions of the following form

$$D^i(\mathbf{p}) = \alpha_i \prod_{j=1}^{n} p_j^{\beta_{ij}}, \tag{6.23}$$

where $\beta_{ii} < -1$ for all i ensures the existence of a Bertrand-Nash equilibrium, which is

$$p_i = \frac{\beta_{ii}}{\beta_{ii}+1}c_i, \quad i = 1,\ldots,n. \tag{6.24}$$

This equilibrium does not depend on other parameters than β_{ii} and c_i. Now consider the adjustment process (6.15). This can be written as

$$p_{i,t+1} = \frac{1}{2}(p_{it} + c_i) - \frac{1}{2}\frac{\alpha_i \prod_{j=1}^{n} p_{jt}^{\beta_{ij}}}{\beta_{ii}p_{it}^{-1}\alpha_i \prod_{j=1}^{n} p_{jt}^{\beta_{ij}}} = \frac{1}{2}\left(1 - \frac{1}{\beta_{ii}}\right)p_{it} + \frac{1}{2}c_i. \tag{6.25}$$

Notice that the dynamics of price of commodity i is independent of the dynamics of the other prices. The stability condition is $\beta_{ii} < -1$ which is always satisfied. Therefore the system with loglinear demand curves and constant marginal costs is always globally stable. This is an example of convergence without specific knowledge of the demand curve.

6.3.2 Example: nonconvergence

Above we have derived conditions under which the Bertrand-Nash equilibrium is locally stable with respect to the short memory learning procedure. In this section we study a specific example of the learning process in which other phenomena than convergence may occur. We assume $n = 2$ and $c_i = 0$ for $i = 1, 2$. Demand functions are given by

$$\begin{aligned} D^1(p_1, p_2) &= \max\left\{\alpha_1 + \beta_{11}p_1^{\gamma_{11}} + \beta_{12}p_2^{\gamma_{12}}, 0\right\}, \\ D^2(p_1, p_2) &= \max\left\{\alpha_2 + \beta_{21}p_1^{\gamma_{21}} + \beta_{22}p_2^{\gamma_{22}}, 0\right\}, \end{aligned} \tag{6.26}$$

where we assume $\beta_{11}, \beta_{22} < 0$ and $\gamma_{11}, \gamma_{22} > 0$. If $\beta_{12}\gamma_{12} > 0$ and $\beta_{21}\gamma_{21} > 0$ the two commodities are substitutes and if $\beta_{12}\gamma_{12} < 0$ and $\beta_{21}\gamma_{21} < 0$ the two commodities are complements. The optimization problem for agent 1 is

$$\max_{p_1} p_1 D^1(p_1, p_2) = \max_{p_1} p_1\left(\alpha_1 + \beta_{11}p_1^{\gamma_{11}} + \beta_{12}p_2^{\gamma_{12}}\right),$$

which gives first order condition

$$\alpha_1 + \beta_{11}(\gamma_{11} + 1)p_1^{\gamma_{11}} + \beta_{12}p_2^{\gamma_{12}} = 0.$$

The second order condition is $\beta_{11}\gamma_{11}(\gamma_{11} + 1)p_i^{\gamma_{11}-1} < 0$, which by assumption is satisfied. Similar first and second order conditions can be derived for firm 2. To simplify the analysis we want $(p_1^*, p_2^*) = (1, 1)$ to be a Bertrand-Nash equilibrium. It can be easily checked that this is indeed an equilibrium if

$$\alpha_1 = \widehat{\alpha}_1 = -\beta_{11}(\gamma_{11} + 1) - \beta_{12} \text{ and } \alpha_2 = \widehat{\alpha}_2 = -\beta_{22}(\gamma_{22} + 1) - \beta_{21}.$$

Notice that the Bertrand-Nash equilibrium does not have to be unique. The dynamics of the learning process (6.15) becomes

$$\begin{aligned} p_{1t+1} &= \frac{1}{2}\left(1 - \frac{1}{\gamma_{11}}\right)p_{1t} - \frac{1}{2}\frac{\widehat{\alpha}_1 + \beta_{12}p_{2t}^{\gamma_{12}}}{\gamma_{11}\beta_{11}p_{1t}^{\gamma_{11}-1}}, \\ p_{2t+1} &= \frac{1}{2}\left(1 - \frac{1}{\gamma_{22}}\right)p_{2t} - \frac{1}{2}\frac{\widehat{\alpha}_2 + \beta_{21}p_{1t}^{\gamma_{21}}}{\gamma_{22}\beta_{22}p_{2t}^{\gamma_{22}-1}}. \end{aligned} \tag{6.27}$$

In order to determine the local stability of $(p_1^*, p_2^*) = (1,1)$ we have to determine the Jacobian evaluated in the Bertrand-Nash equilibrium, which turns out to be

$$\mathbf{J} = \begin{pmatrix} \frac{1}{2} - \frac{1}{2}\gamma_{11} & -\frac{1}{2}\frac{\gamma_{12}\beta_{12}}{\gamma_{11}\beta_{11}} \\ -\frac{1}{2}\frac{\gamma_{21}\beta_{21}}{\gamma_{22}\beta_{22}} & \frac{1}{2} - \frac{1}{2}\gamma_{22} \end{pmatrix}.$$

The eigenvalues of this Jacobian matrix are

$$\lambda_{1,2} = \frac{1}{4}\left((2 - \gamma_{11} - \gamma_{22}) \pm \sqrt{(\gamma_{11} - \gamma_{22})^2 + 4\frac{\gamma_{12}\beta_{12}\gamma_{21}\beta_{21}}{\gamma_{11}\beta_{11}\gamma_{22}\beta_{22}}}\right) \quad (6.28)$$

Therefore a necessary, but not sufficient, condition for these eigenvalues to be complex is $(\gamma_{12}\beta_{12}) \times (\gamma_{21}\beta_{21}) < 0$, that is, one of the commodities is a substitute to the other commodity, while the latter is a complement to the former. This seems to be an unrealistic case. However, since we know that any set of continuous functions can be the aggregate excess demand functions it is possible that it occurs. For the sake of completeness, the next result states what might happen then. Notice, however, that these phenomena are not so peculiar in examples with three or more different firms and commodities.

Proposition 6.6 *Assume* $\gamma_{11} = \gamma_{22} = \gamma > 0$, $\beta_{11} = \beta_{22} = -1$, $\beta_{12} = \beta_{21} = \beta > 0$, *and* $\gamma_{12} = -\gamma_{21} = \delta$. *If* $\gamma > 3$ *the symmetric equilibrium* $(p_1^*, p_2^*) = (1,1)$ *is unstable for all values of* β *and* δ. *If* $\gamma \in [0,3]$ *then the equilibrium* (p_1^*, p_2^*) *is stable (unstable) for* $\delta < (>)\, \delta^*$, *where*

$$\delta^* = \frac{\gamma}{\beta}\sqrt{3 + 2\gamma - \gamma^2}.$$

At this value for δ *a Hopf bifurcation occurs and an invariant circle emerges.*

Proof. The eigenvalues are equal to

$$\lambda_{1,2} = \frac{1}{2}\left(1 - \gamma \pm i\frac{\beta\delta}{\gamma}\right).$$

Therefore if $\gamma > 3$, both eigenvalues lie outside the unit circle. If $\gamma < 3$ then the absolute value of both eigenvalues is equal to their product and hence

$$\lambda_1 \times \lambda_2 = \frac{1}{4}(1 - \gamma)^2 - \frac{1}{4}\left(\frac{\beta\delta}{\gamma}\right)^2.$$

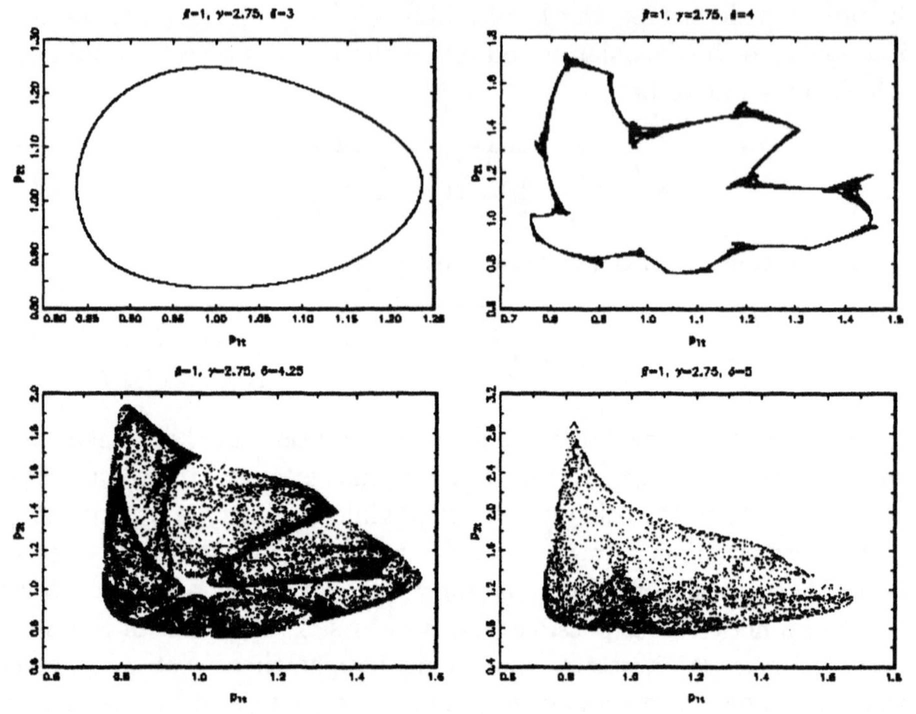

Figure 6.1: Attractors of the short memory learning model with $\beta_{11} = \beta_{22} = -1$, $\beta_{12} = \beta_{21} = 1$, $\gamma_{11} = \gamma_{22} = 2\frac{3}{4}$, $\alpha_1 = \alpha_2 = 3\frac{3}{4}$ and $\gamma_{12} = -\gamma_{21} = \delta$, where we have: a) $\delta = 3$, b) $\delta = 4$, c) $\delta = 4\frac{1}{4}$ and d) $\delta = 5$.

This product is equal to 1 when $\delta = \delta^*$ and therefore at δ^* the eigenvalues cross the unit circle. ■

At $\delta = \delta^*$ an invariant circle emerges. For higher values of δ even more complicated attractors exist. Figure 6.1 gives examples of these attractors, for $\beta = 1$ and $\gamma = 2\frac{3}{4}$. A Hopf bifurcation occurs at $\delta^* = \frac{11}{16}\sqrt{15} \approx 2.66$. For values of $\delta > \delta^*$ an attracting invariant cycle exists. This invariant cycle is shown for $\delta = 3$. More complicated attractors are shown for $\delta = 4$, $\delta = 4\frac{1}{4}$ and for $\delta = 5$.

Now focus on the case where both commodities are substitutes *or* both are complements.

Proposition 6.7 *Assume* $\gamma_{11} = \gamma_{22} = \gamma > 0$, $\beta_{11} = \beta_{22} = -1$, $\beta_{12} = \beta_{21} = \beta$, *and* $\gamma_{12} = \gamma_{21} = \delta$. *Then the eigenvalues are real and we have the following*

1. *If* $\gamma > 3$ *the equilibrium* $(p_1^*, p_2^*) = (1,1)$ *is unstable for all values of* δ *and* β,

2. *if* $1 < \gamma < 3$, *the equilibrium* $(p_1^*, p_2^*) = (1,1)$ *is stable (unstable) for* $|\delta| < (>) |\delta^*|$, *where*

$$|\delta^*| = \frac{\gamma}{\beta}(3-\gamma).$$

At this value of δ *a period doubling bifurcationoccurs and a period two orbit emerges.*

3. *If* $0 < \gamma < 1$, *the equilibrium* $(p_1^*, p_2^*) = (1,1)$ *is stable (unstable) for* $|\delta| < (>) |\delta^0|$, *where*

$$|\delta^0| = \frac{\gamma}{\beta}(1+\gamma).$$

Proof. The eigenvalues are

$$\lambda_{1,2} = \frac{1}{2}\left(1-\gamma \pm \frac{\beta\delta}{\gamma}\right).$$

Notice that for $\gamma > 3$, there is always at least one eigenvalue smaller than -1 implying that the equilibrium is unstable. Without loss of generality assume $\delta > 0$. Then the largest eigenvalue corresponds to the positive root. We have to determine when this eigenvalue is smaller than 1. This gives

$$\frac{1}{2}\left(1-\gamma+\frac{\beta\delta}{\gamma}\right) \leq 1 \Rightarrow \delta \leq \delta^0 = \frac{\gamma}{\beta}(1+\gamma)$$

The smallest eigenvalue corresponds to the negative root. We have to determine when it is bigger than -1. This gives

$$\frac{1}{2}\left(1-\gamma-\frac{\beta\delta}{\gamma}\right) \geq -1 \Rightarrow \delta \leq \delta^* = \frac{\gamma}{\beta}(3-\gamma)$$

Clearly $\delta^0 < (>) \delta^*$ if $\gamma < (>) 1$. Therefore if $\gamma > 1$, the equilibrium loses stability through a period doubling bifurcation at $\delta = \delta^*$ and if $\gamma < 1$ the equilibrium loses stability either through a transcritical or through pitchfork bifurcation at $\delta = \delta^0$. Whether a transcritical or a pitchfork bifurcation occurs depends upon the parameters of the model. ■

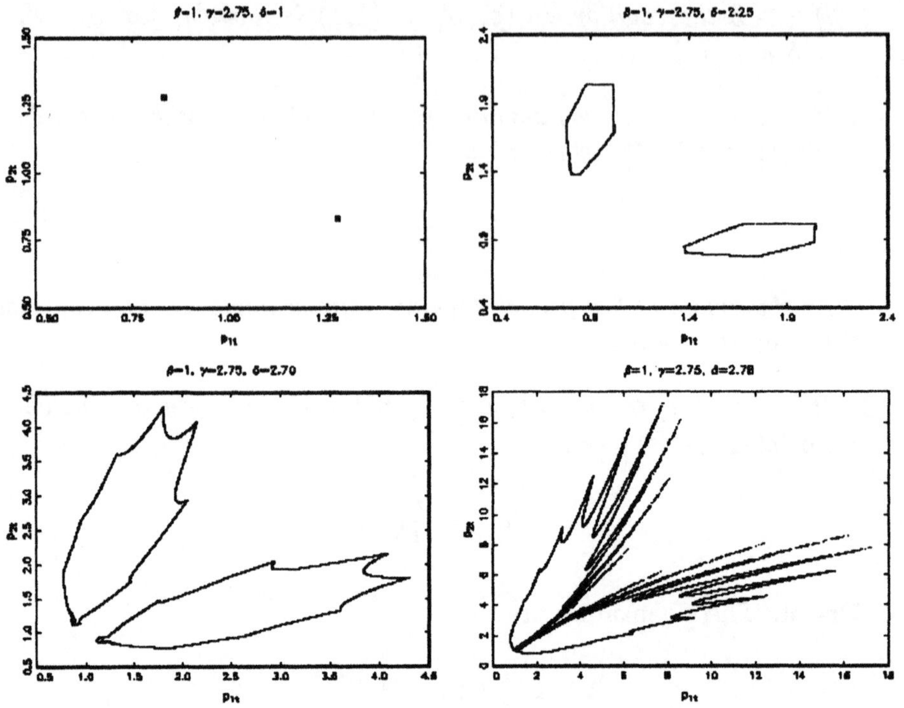

Figure 6.2: Attractors of the short memory learning model with $\beta_{11} = \beta_{22} = -1$, $\beta_{12} = \beta_{21} = 1$, $\gamma_{11} = \gamma_{22} = 2\frac{3}{4}$, $\alpha_1 = \alpha_2 = 3\frac{3}{4}$ and $\gamma_{12} = \gamma_{21} = \delta$, where we have: a) $\delta = 1$, b) $\delta = 2\frac{1}{4}$, c) $\delta = 2.70$ and d) $\delta = 2.78$.

Notice that $\gamma < 1$, implies that the demand for commodity i is strictly concave in p_i and $\gamma > 1$ implies that demand for commodity i is strictly convex in p_i.

Now consider the case $1 < \gamma < 3$. There is an important difference between the dynamics of the case with substitutes and the case with complements. If the commodities are substitutes, this leads to an attractor which lies off the diagonal $y = x$, but is symmetric with respect to this diagonal. This in general means that the following occurs: when one price is low, the other is high and vice versa. If the commodities are complements, the attractor lies along the $y = x$ axis and prices move together: both are low or both are high.

As example we take the following parameter setting: $\beta = 1$, $\gamma = 2\frac{3}{4}$ and $\delta > 0$. Figure 6.2 shows some attractors for this case. A period doubling bifurcation occurs at $\delta^* = \frac{11}{16} = 0.6875$. The resulting period two

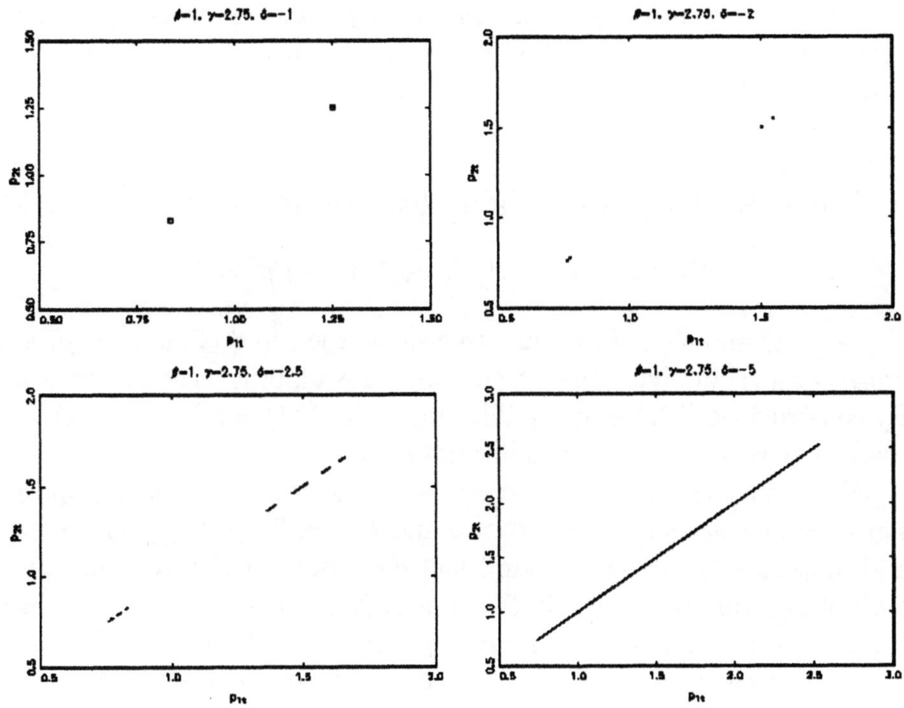

Figure 6.3: Attractors of the short memory learning model with $\beta_{11} = \beta_{22} = -1$, $\beta_{12} = \beta_{21} = 1$, $\gamma_{11} = \gamma_{22} = 2\frac{3}{4}$, $\alpha_1 = \alpha_2 = 3\frac{3}{4}$ and $\gamma_{12} = \gamma_{21} = \delta$, where we have: a) $\delta = -1$, b) $\delta = -2$, c) $\delta = -2\frac{1}{2}$ and d) $\delta = -5$.

orbit undergoes a Hopf bifurcation, leading into a invariant set consisting of two closed curves. This invariant set then evolves into a strange attractor as shown in Figure 6.2. As another example, take the case where the commodities are complements: $\beta = 1$, $\gamma = 2\frac{3}{4}$ and $\delta < 0$. A period doubling bifurcation occurs again at $\delta^* = \frac{11}{16}$. This bifurcation is followed by a sequence a period doubling bifurcations on the $y = x$ axis, resulting in a one dimensional strange attractor as shown in Figure 6.3.

We have throughout concentrated on the symmetric equilibrium, however it is clear from the last result that multiple equilibria may exist. The following result states when multiple symmetric equilibria exist in the symmetric version of the model.

Proposition 6.8 *Assume* $\gamma_{11} = \gamma_{22} = \gamma > 0$, $\beta_{11} = \beta_{22} = -1$, $\beta_{12} = \beta_{21} = \beta$, *and* $\gamma_{12} = \gamma_{21} = \delta > 0$, *with* $\delta \neq \gamma$. *Consider* $A = (\gamma + 1 - \beta) \times$

$(\delta - \gamma)$. *If* $A > 0$ *a sufficient conditions for the existence of multiple symmetric equilibria is* $\delta\beta \neq \gamma(\gamma + 1)$. *If* $A < 0$ *and* $\delta < \gamma$ *a sufficient condition is* $\delta\beta < \gamma(\gamma + 1)$, *if* $A < 0$ *and* $\delta > \gamma$ *a sufficient condition is* $\delta\beta > \gamma(\gamma + 1)$.

Proof. A symmetric equilibrium (p, p) has to satisfy

$$F(p) = \gamma + 1 - \beta - (\gamma + 1) p^{\gamma} + \beta p^{\delta} = 0.$$

If $\gamma + 1 - \beta$ and $\delta - \gamma$ have the same signs, then so do $\lim_{p \to 0} F(p)$ and $\lim_{p \to \infty} F(p)$, implying that the number of equilibria is generically even. By construction $(1, 1)$ is an equilibrium, so if $F'(1) = \delta\beta - \gamma(\gamma + 1) \neq 0$, there is at least one other equilibrium $(\widehat{p}, \widehat{p})$.

Now consider the case where $\gamma + 1 - \beta$ and $\delta - \gamma$ have opposite signs. First consider $\gamma + 1 - \beta > 0$ and $\delta < \gamma$. Then $\lim_{p \to 0} F(p) > 0$ and $\lim_{p \to \infty} F(p) = -\infty$. A sufficient condition for the existence of at least three equilibria, then is $F'(1) > 0$. Next consider the case where $\gamma + 1 - \beta < 0$ and $\delta > \gamma$. Then $\lim_{p \to 0} F(p) < 0$ and $\lim_{p \to \infty} F(p) = \infty$. A sufficient condition for the existence of at least three equilibria now is $F'(1) < 0$. ∎

What is the economic interpretation of the nonconvergence results established above? Clearly, when the system converges to a periodic orbit with a given period k, then agents can predict demand in each period much better by looking at past data. Therefore, it might then not be too sensible to follow the procedure discussed here. However, if the time series of prices show no clear pattern then these fluctuations might be attributed to stochastic demand shocks, and this may leave no reason for a single firm to doubt his specification of the demand structure.

6.4 A long memory learning procedure

We discussed three adjustment models for our heterogeneous Bertrand oligopoly, the best-reply dynamics, gradient systems and the short memory learning procedure discussed in Section 6.3. All these adjustment processes only use the most recent information (with different degrees of sophistication) about competitors actions or price-quantity combinations. We have seen that this may lead to nonconvergent behaviour such as cycles and even chaotic fluctuations. Clearly, when agents observe that they are in a cycle they can use this knowledge to improve their predictions. These new predictions will again have its ramifications on the behaviour of the system which may lead to, for example, a cycle with a different period altogether. An alternative approach would be to consider a learning procedure where firms use information on the total history of price-quantity pairs available to make inferences on their perceived demand structures. In this section we will review the model that was studied by Kirman in a series of papers (Kirman (1975, 1983, 1995) and Brousseau and Kirman (1992, 1993)). An interpretation and generalisation of this learning process in terms of Bayesian learning can be found in Schinkel, Tuinstra and Vermeulen (1999), but it would take us too far to discuss that paper in detail here. In the model by Kirman a symmetric price-setting duopoly model with linear objective demand curves is considered where the two firms estimate a misspecified model by running a least squares regression on past quantities and *own* prices. Simulation results suggest that this learning procedure converges to the set of *conjectural* equilibria, to be defined later, but a theoretical result establishing this convergence has not been provided yet.

Before we turn to the Kirman model we first mention a related line of research by Gates, Rickard and Wilson (1977, 1978). They consider a Cournot oligopoly model where firms use a weighted least squares procedure to estimate a linear relationship between output and average profit (that is profit per unit output). Output is then adjusted in the direction of the output that maximizes the estimated profit function. As in the Kirman model Gates, Rickard and Wilson (1977) use simulations to show that there is convergence. For the case of a duopoly they explicitly calculate the set of limit points. Gates, Rickard and Wilson (1978) consider a special case of this model, where only the first and last observation are used. They are able to prove that in this case, if average profit functions are linear in output, the process converges.

Kirman (1975, 1983) considers the following symmetric duopoly model with heterogenous commodities

$$\begin{aligned} D^1(p_1, p_2) &= \alpha - \beta p_1 + \gamma p_2, \\ D^2(p_1, p_2) &= \alpha - \beta p_2 + \gamma p_1. \end{aligned} \tag{6.29}$$

We assume zero costs, implying that firms want to maximize total revenue. The Bertrand Nash equilibrium is $p_1^* = p_2^* = \frac{\alpha}{2\beta - \gamma}$ and the cartel solution, where agents maximize their joint profit, is $p_1^c = p_2^c = \frac{\alpha}{2(\beta-\gamma)}$ (provided of course, that these prices are positive). Firm i believes that demand only depends upon its own price and therefore

$$\begin{aligned} d_1(p_1) &= a_1 - b_1 p_1 + \varepsilon_1, \\ d_2(p_2) &= a_2 - b_2 p_2 + \varepsilon_2, \end{aligned} \tag{6.30}$$

respectively, where they believe $\varepsilon_i \sim N\left(0, \sigma_i^2\right)$ makes up appropriately for the empirical deviations of quantity signals from expected sales.

The parameters a_1, b_1, a_2, and b_2 are estimated in each period t as a_{1t}, b_{1t}, a_{2t} and b_{2t} respectively, by applying an ordinary least square regression on all past observations of (self-made) prices and associated demand signals. Kirman (1975) shows that if producers know the slope β of the individual demand curve and only have to estimate the intercept of the demand curve prices will converge to the Bertrand-Nash equilibrium. Formally this version of the model is equivalent with fictitious play, where each producer has a correctly specified model but has to make some inference about the price of its competitor and does this on the basis of the average of this price.

We follow Kirman (1983) in assuming that the slope of the demand function is also unknown. We then define a *conjectural equilibrium* as a vector of belief parameters $\left(\widehat{a}_1, \widehat{b}_1, \widehat{a}_2, \widehat{b}_2\right)$ and a pair of prices $(\widehat{p}_1, \widehat{p}_2)$ such that firms' expectations with respect to supply possibilities are satisfied. This occurs when objective demand equals (expected) perceived demand, that is

$$\begin{aligned} D^1(p_1, p_2) &= \alpha - \beta\widehat{p}_1 + \gamma\widehat{p}_2 = \widehat{a}_1 - \widehat{b}_1\widehat{p}_1 = d_1(\widehat{p}_1) \\ D^2(p_1, p_2) &= \alpha - \beta\widehat{p}_2 + \gamma\widehat{p}_1 = \widehat{a}_2 - \widehat{b}_2\widehat{p}_2 = d_2(\widehat{p}_2) \end{aligned}$$

where the prices have to be such that firms believe that they maximize profits, hence

$$\widehat{p}_1 = \frac{\widehat{a}_1}{2\widehat{b}_1} \text{ and } \widehat{p}_2 = \frac{\widehat{a}_2}{2\widehat{b}_2}.$$

The set of conjectural equilibria is very large, in fact it can easily be seen that any pair of prices (p_1, p_2) for which demand is positive can be a conjectural equilibrium, just by taking

$$b_1 = \frac{\alpha - \beta p_1 + \gamma p_2}{p_1}, \quad b_2 = \frac{\alpha - \beta p_1 + \gamma p_2}{p_1}, \ a_1 = 2b_1 p_1 \text{ and } a_2 = 2b_2 p_2.$$

The least squares regression referred to above gives the following estimates

$$b_{i,t+1} = -\frac{\sum_{k=1}^{t}\left[d_{ik} - \overline{d}_{it}\right]\left[p_{ik} - \overline{p}_{it}\right]}{\sum_{k=1}^{t}\left[p_{ik} - \overline{p}_{it}\right]^2} \text{ and } a_{it} = \overline{d}_{it} + b_{it}\overline{p}_{it},$$

where $\overline{d}_{it} = \frac{1}{t}\sum_{k=1}^{t} d_{ik}$ and $\overline{p}_{it} = \frac{1}{t}\sum_{k=1}^{t} p_{ik}$ are average demand signals and average prices at time t respectively. It is well known that the ordinary least squares regression can also be written recursively (see also the recursive least squares learning models in Chapter 4). Define

$$\theta_{it} = \begin{pmatrix} a_{it} \\ b_{it} \end{pmatrix} \text{ and } x_{it} = \begin{pmatrix} 1 & -p_{it} \end{pmatrix},$$

then the estimate at time t using ordinary least squares becomes

$$\theta_{it} = \left(X'_{it}X_{it}\right)^{-1} X'_{it}D_{it}$$

where

$$X_{it} = \begin{pmatrix} 1 & -p_{i1} \\ \vdots & \vdots \\ 1 & -p_{it} \end{pmatrix} \text{ and } D_{it} = \begin{pmatrix} d_{i1} \\ \vdots \\ d_{it} \end{pmatrix}.$$

Now define $R_{it} \equiv \frac{1}{t}X'_{it}X_{it}$. We then obtain

$$\begin{aligned} R_{it} &= \frac{1}{t}X'_{it}X_{it} = \frac{1}{t}\left(X'_{i,t-1}X_{i,t-1} + x'_{it}x_{it}\right) \\ &= \frac{1}{t-1}X'_{i,t-1}X_{i,t-1} + \frac{1}{t}\left(x'_{it}x_{it} - \frac{1}{t-1}X'_{i,t-1}X_{i,t-1}\right) \\ &= R_{i,t-1} + \frac{1}{t}\left(x'_{it}x_{it} - R_{i,t-1}\right) \end{aligned}$$

The estimate θ_{it} can be written recursively since.

$$\begin{aligned} \theta_{it} &= \left(X'_{it}X_{it}\right)^{-1} X'_{it}D_{it} \\ &= \left(X'_{it}X_{it}\right)^{-1}\left[X'_{i,t-1}D_{i,t-1} + x'_{it}d_{it}\right] \\ &= \left(X'_{it}X_{it}\right)^{-1}\left[X'_{i,t-1}X_{i,t-1}\theta_{i,t-1} + x'_{it}d_{it}\right] \\ &= \left(X'_{it}X_{it}\right)^{-1}\left[\left(X'_{it}X_{it} - x'_{it}x_{it}\right)\theta_{i,t-1} + x'_{it}d_{it}\right] \\ &= \theta_{i,t-1} + \left(X'_{it}X_{it}\right)^{-1} x'_{it}\left(d_{it} - x_{it}\theta_{i,t-1}\right). \end{aligned}$$

The whole system therefore becomes

$$\begin{aligned} R_{it} &= R_{i,t-1} + \frac{1}{t}\left(x'_{it}x_{it} - R_{i,t-1}\right) \qquad (6.31) \\ \theta_{it} &= \theta_{i,t-1} + \frac{1}{t}R_{it}^{-1}x'_{it}\left(d_{it} - x_{it}\theta_{i,t-1}\right) \end{aligned}$$

Starting with the right initial condition θ_{i0} and R_{i0} this gives the exact same estimates as the non-recursive least squares procedure.

Before agents can estimate a curve, at least two price-demand combinations have to be known, therefore the process needs some initial conditions. If these initial conditions are the same for both firms, that is, $p_{1t} = p_{2t}$ for $t = 1, 2$, then the price quantity pairs of the first two periods lie on the demand curve

$$d = \alpha - (\beta - \gamma) p$$

and therefore prices will be equal to the cartel solution $p^c = \alpha / (2\beta - 2\gamma)$ from the third period on. Kirman (1983) furthermore shows that if a conjectural equilibrium $\left(\widehat{a}_1, \widehat{b}_1, \widehat{a}_2, \widehat{b}_2\right)$ satisfies the following inequalities

$$\gamma^2 \geq \left(\beta - \widehat{b}_1\right)\left(\beta - \widehat{b}_2\right) > 0,$$

there exist initial conditions for the first three periods, such that the learning procedure will jump to this conjectural equilibrium in the fourth period, and stay there. He calls these equilibria *sustainable equilibria* since they are possible limit points of the learning algorithm. Of course we are mainly interested in what happens for arbitrary initial conditions instead of what happens for specific ones. Simulations suggest that the least squares learning procedure always converges to a conjectural equilibrium. We have also performed some simulations for some nonlinear objective demand functions and then the least squares learning procedure converges as well, so the result seems to hold for a wide variety of specifications. Notice that for these models the learning process hardly restricts the set of possible equilibria.

In the literature there are two types of convergence results for these kinds of recursive least squares algorithms, which we will briefly discuss in the following.

First focus on the recursive formulation (6.31) of the ordinary least squares algorithm. Convergence of these types of algorithms can be shown by using a technique developed by Ljung (see Ljung (1977) and Ljung and Söderström (1983)). For applications of this technique in economics we refer to Marcet and Sargent (1989), Sargent (1993), Woodford (1990) and Evans and Honkahpohja (1997). The idea is to construct a (deterministic and autonomous) associated differential equation, which is related to the original (stochastic and nonautonomous) recursive algorithm. Under some regularity conditions stability of the equilibria of this associated difference equation implies convergence (almost surely) of the recursive algorithm itself.

Brousseau and Kirman (1992) apply this procedure to the symmetric duopoly model set out above. They find that none of the conjectural equilibria are locally stable. This is due to the fact that there is a continuum of equilibria which implies that a slight perturbation in initial conditions leads the learning procedure to converge to a different equilibrium. convergence will be to a different equilibrium. Therefore there seems to be global convergence on the set of equilibria, but none of these equilibria is locally stable. Kirman (1995) holds that this is a case of 'apparent' convergence in the sense that the estimates keep on changing only at a rate that slows down until these adjustments are not observable anymore. This could be due to the fact that new information is weighed less and less as time goes by (as can be seen from the term $1/t$ in (6.31)). Time averages need not converge. We saw that this could also happen with some of the recursive least squares learning models in Chapter 4.[6] The least squares algorithm set out above is a so-called *decreasing gain* algorithm: the gain from a new observation is decreasing since all observations get the same weight. An alternative would be to consider so-called *constant gain* algorithms.[7] These algorithms have the property that the gain of new observations remains constant because these observations are weighed more heavily than older ones. For (6.31) this just means replacing $1/t$ by a constant. We have performed simulations in the Kirman model with this constant gain algorithm and all these simulations have lead to convergence to a conjectural equilibrium, as was the case with the decreasing gain algorithm. This suggests that, in contrast to Kirman's conjecture, true convergence may occur, even in the case of the decreasing gain algorithm.

Another route to establish convergence is to consider Bayesian learning. In the last couple of decades there has been an increasing number of applications of Bayesian learning to economic models. Blume and Easley (1982, 1984), Bray and Savin (1986) and Bray and Kreps (1987) have, in different contexts, investigated whether Bayesian learning leads to a rational expectations equilibrium or not. Easley and Kiefer (1988), Kiefer and Nyarko (1989), Nyarko (1991) and others have addressed the issue of "active learning". They study a monopolist who does not know the

[6] Other examples of this phenomenon can be found in Shapley (1964) and Gaunersdorfer and Hofbauer (1995).

[7] For an interesting example of an application of a constant gain algorithm see Sargent (1998).

demand curve he faces and who wants to maximize a discounted stream of profits. Actively learning firms are aware of the fact that there is a trade-off between learning and short-run profit maximization. For them it might be optimal to forfeit current profits to experiment with prices in order to gain information about the location of the demand curve, which may lead to higher future profits. Jordan (1991), Kalai and Lehrer (1993, 1995) and Nyarko (1994) have focussed on the conditions under which Bayesian learning leads to a Nash equilibrium in normal form games.

Models of Bayesian learning assume that agents have some prior distribution over the parameters they want to estimate. Let us denote this prior distribution for firm i by $\mu_{i0}(\alpha_i, \beta_i)$. Given this prior distribution firms choose a price that optimizes their expected profit. This leads to new information about the quantity that can be sold at that price. The beliefs of the firms are then updated according to Bayes' rule. These new beliefs $\mu_{i1}(\alpha_i, \beta_i)$ for firm i lead to a new price again, and this process repeats over and over. If beliefs $\mu_{it}(\alpha_i, \beta_i)$ converge then so do prices and quantities. Most of the literature on Bayesian' learning deals with well specified models, in which the prior distribution over the different models gives positive weight to the true model. We are interested in mis-specified models, where the true model cannot be learned by the agents. It is possible, however, that their beliefs converge and that, in the end, the firms receive no information that makes them want to change their distribution over different models. To prove that Bayesian' learning converges one proceeds in the following way. First, it is shown that the belief μ_{it} is a uniformly bounded *martingale*, that is, the expectation of agent i of the belief $\mu_{i,t+1}$ at time t equals μ_{it}. Then, by the martingale convergence theorem, beliefs μ_{it} converge almost surely, that is beliefs converge for almost all time paths of prices and quantities that are possible according to the prior distribution μ_{i0}. In other words, for all sets of realizations of the trajectories of prices and quantities that have been given positive probability at time 0 in the minds of the firms, the Bayesian' learning model converges to some limit belief. This could be called *subjective convergence of beliefs*. Then the final step to relate this subjective convergence to objective convergence is to assume that the objective probability measure of all possible objective sample paths is *absolutely continuous* with respect to the subjective probability of all sample paths, that is, each set of sample paths that has positive probability objectively has positive probability subjectively. Absolute continuity implies that no actual development is possible that was not *a priori* foreseen as a possibility by the firm concerned. This seems to be a natural condition necessary for beliefs to settle down, as one can hardly expect beliefs to converge if all the time new and unforeseen events stir up ones learning.

The Kirman model can be interpreted as a model of Bayesian learning. Suppose both sellers have a normal prior $\mu_{i0}(\theta_i)$, that is, agent i beliefs that θ_i is distributed according to $N(\theta_{i0}, \Sigma_{i0})$, where $\theta_{i0} = (a_{i0}, b_{i0})'$ is a 2×1 vector, and Σ_{i0} a 2×2 finite and nonsingular variance-covariance matrix. Furthermore, sellers believe θ_i is uncorrelated with the error term for all t. In this situation Bayesian learning can be shown to be equivalent with weighted least squares. Therefore convergence of Bayesian learning implies convergence of least-squares learning in the Kirman model. For more details on Bayesian learning in mis-specified models see Schinkel, Tuinstra and Vermeulen (1999).

6.5 Concluding remarks

We have been addressing one of the unresolved problems of economic theory, the problem of adjustment toward equilibrium. Clearly if this problem cannot be dealt with satisfactorily, the focus of the economics profession on the study of equilibria might be unwarranted. Traditional adjustment processes such as the tâtonnement process are unsatisfactory for two reasons. Firstly, they are unrealistic and secondly, the conditions under which convergence to an equilibrium indeed occurs are too special. It can be argued that a more realistic adjustment process would incorporate agents that set prices themselves. In this chapter we have investigated some simple adjustment processes in equilibrium models with price setting agents. These adjustment processes share the feature that firms do not take the behaviour of other firms into account when determining their optimal strategy. All firms act as if they are monopolists. Furthermore they have incomplete knowledge of the demand curve they face. Therefore, the models of the economy these firms have are misspecified. In Chapter 4 it was already argued that this seems to be most sensible way to approach learning. The adjustment processes differ in the way in which inferences are made about this demand curve. One process uses recent information about the amount that could be sold against yesterday's price and the sensitivity of demand at yesterday's price (for example obtained by small price experiments) to estimate a linear demand curve, the other process uses all previous experiences on quantity-price combinations in order to estimate a linear demand curve using standard regression techniques. The first adjustment process appears to be stable for situations where the cross price effects are relatively small. This seems to be in accordance with the assumption that firms act monopolistically. The second adjustment process grants convergence to a conjectural equilibrium in most situations. Thus far, these adjustment models have only been studied in a partial equilibrium context. These models should also be applied to general equilibrium models of monopolistic competition.

The analysis could be extended in a number of other directions. As mentioned before any realistic adjustment process should allow for the possibility of trade at disequilibrium prices. Furthermore, we have focused on myopic optimizers. Even though they sometimes use a lot of information to estimate their profit opportunities, firms are only interested in short term profit maximization. Ideally firms would try to maximize some discounted stream of profits, but this complicates the problem considerably. Every firm then has to solve an optimal control problem which takes account of the trade off between the benefits of experimenting with prices in order to gain information on the demand curve and the resulting loss in short run profits. Clearly, much work remains to be done on the subject of price adjustment processes.

References

ANDERSON, S.P., A. DE PALMA AND J.F. THISSE. (1992): *Discrete Choice Theory of Product Differentiation.* Cambridge, MA: MIT Press.

ARIFOVIC, J. (1995): Genetic algorithms and inflationary economies. *Journal of Monetary Economics,* **36,** 219-243.

ARIFOVIC, J. (1996): The behavior of the exchange rate in the genetic algorithm and experimental economics. *Journal of Political Economy,* **104,** 510-541.

ARROW, K.J. (1959): Towards a theory of price adjustment. In M. Abramovitz et al: *the Allocation of Economic Resources,* Stanford: Stanford University Press.

ARROW, K.J. AND F.H. HAHN (1971): *General Competitive Analysis,* San Francisco: Holden-Day.

ARROW, K.J., H.D. BLOCK AND L. HURWICZ. (1959): On the stability of the competitive equilibrium, II. *Econometrica,* **27,** 82-109.

ARROW, K.J. AND L. HURWICZ. (1960): Stability of the gradient process in n-person games. *Journal of the Society for Industrial and Applied Mathematics,* **8,** 280–294.

ATKINSON, K.E. (1989): *An Introduction to Numerical Analysis.* New York: John Wiley & Sons.

AXELROD, R.M. (1984): *The Evolution of Cooperation.* New York: Basic Books.

BALA, V. (1997): A pitchfork bifurcation in the tatonnement process. *Economic Theory,* **10,** 521-530.

BALA, V. AND M. MAJUMDAR (1992): Chaotic tâtonnement. *Economic Theory,* **2,** 437-445.

BALASKO, Y AND C. GHIGLINO (1995): On the existence of endogenous cycles. *Journal of Economic Theory,* **67,** 566-577.

BANERJEE, A., AND J.W. WEIBULL (1995): Evolutionary selection and rational behavior. In *Learning and Rationality in Economics* (A. Kirman and M. Salmon, Eds.), pp. 343-363. Oxford: Blackwell Publishers.

BENASSY, J.-P. (1982): *The Economics of Market Disequilibrium*, Academic Press, New York.

BENASSY, J.-P. AND M.C. BLAD (1989): On learning and rational expectations in an overlapping generations model. *Journal of Economic Dynamics and Control*, **13**, 379-400.

BENEDICKS, M. AND L. CARLESON (1991): The dynamics of the Hénon map. *Annals of Mathematics,* **133**, 73-169.

BENHABIB, J. AND R.H. DAY (1982): A characterization of erratic dynamics in the overlapping generations model. *Journal of Economic Dynamics and Control,* **4**, 37-55.

BINMORE, K. AND L. SAMUELSON (1997): Muddling through: noisy equilibrium selection. *Journal of Economic Theory,* **74**, 235-265.

BISCHI, G.-I. AND L. GARDINI (1997): Cycles and bifurcations in duopoly games. Working Paper University of Urbino.

BISCHI, G.-I. AND A. NAIMZADA (1997): A dynamic duopoly game with local adaptive response. Working Paper University of Urbino.

BISCHI, G-.I., L. GARDINI AND M. KOPEL (2000). Analysis of global bifurcations in a market share attraction model. *Journal of Economic Dynamics and Control,* **24**, 855-879.

BISCHI, G-.I., L. GARDINI AND C. MIRA (1999). Plane maps with denominator I. Some generic properties. *International Journal of Bifurcation and Chaos,* **9**, 119-153.

BJÖRNERSTEDT, J. K.H. AND SCHLAG (1994): The evolution of imitative behavior. Mimeo, Department of Economics, University of Bonn.

BLUME, L. AND D. EASLEY (1982): Learning to be rational. *Journal of Economic Theory,* **26**, 340-351.

BLUME, L. AND D. EASLEY (1984): Rational expectations equilibrium: an alternative approach. *Journal of Economic Theory,* **34**, 116-129.

BÖHM, V. (1994): The foundation of the theory of monopolistic competition revisited. *Journal of Economic Theory,* **63**, 208-218.

BÖHM, V. AND J. WENZELBURGER (1999): Expectations, forecasting and perfect foresight – a dynamical systems approach. *Macroeconomic Dynamics,* **3**, 167-186.

BONANNO, G. (1988): Oligopoly equilibria when firms have local knowledge of demand. *International Economic Review*, **29**, 45-55.

BONANNO, G. (1990): General equilibrium theory with imperfect competition. *Journal of Economic Surveys*, **4**, 297-328.

BONANNO, G. AND E.C. ZEEMAN (1985): Limited knowledge of demand and oligopoly equilibria. *Journal of Economic Theory*, **35**, 276-283.

BRAY, M.M. (1982): Learning, estimation, and the stability of rational expectations. *Journal of Economic Theory*, **26,** 318-339.

BRAY, M.M. AND N.E. SAVIN (1986): Rational expectations equilibria, learning, and model specification. *Econometrica*, **54**, 1129-1160.

BRAY, M. AND D. KREPS (1987): Rational learning and rational expectations. In: G. Feiwell (ed.), *Arrow and the Ascent of Modern Economic Theory*, New York: New York University Press.

BROCK, W.A. AND C.H. HOMMES (1997): A rational route to randomness. *Econometrica*, **65**, 1059-1095.

BROUSSEAU, V. AND A.P. KIRMAN (1992): Apparent convergence of learning processes in mis-specified games. In: B. Dutta et.al. (eds.), *Game Theory and Economic Applications*, Berlin: Springer-Verlag, Berlin, 1992.

BROUSSEAU, V. AND A.P. KIRMAN (1993): The dynamics of learning in n-person games with the wrong n. In: K. Binmore, A.P. Kirman, and P. Tani (eds.), *Frontiers of Game Theory*, Cambridge, MA: MIT Press.

BULLARD, J. (1994): Learning equilibria. *Journal of Economic Theory,* **64,** 468-485.

BULLARD, J. AND J. DUFFY (1998): On learning and the stability of cycles. *Macroeconomic Dynamics,* **2,** 22-48.

BULLARD, J. AND J. DUFFY (1999): Using genetic algorithms to model the evolution of heterogenous beliefs. *Computational Economics,* **13,** 41-60.

CABRALES, A. AND T. HOSHI (1996): Heterogeneous beliefs, wealth accumulation, and asset price dynamics. *Journal of Economic Dynamics and Control,* **20**, 1073-1100.

CHATTERJI, S. (1995): Temporary equilibrium dynamics with Bayesian learning. *Journal of Economic Theory,* **67,** 590-598.

CHEN, H.-C., J.W. FRIEDMAN AND J.-F. THISSE (1997): Boundedly rational Nash equilibrium: a probabilistic choice approach. *Games and Economic Behavior,* **18**, 32-54.

CHOSSAT, P. AND M. GOLUBITSKY (1988): Symmetry-increasing bifurcation of chaotic attractors. *Physica D*, **32**, 423-436.

CONLISK, J. (1996): Why bounded rationality? *Journal of Economic Literature*, **34**, 669-700.

CORCHÓN, L.C. AND A. MAS-COLLEL (1995): A note on best reply and gradient systems with applications to imperfectly competitive models. Working paper University of Alicante, WP-AD 95-18.

COURNOT, A. (1838): *Recherches sur les Principes Mathématiques de la Théorie des Richesses,* English translation (1927) by N.T. Bacon: *Mathematical Principles of the Theory of Wealth*, New York: Augustus M. Kelly.

COX, J.C., AND M. WALKER (1998): Learning to play Cournot duopoly strategies. *Journal of Economic Behavior & Organization*, **36,** 141-161.

DANA, R.A. AND L. MONTRUCCHIO (1986): Dynamic complexity in duopoly games. *Journal of Economic Theory*, **40**, 40-56.

DAY, R.H. (1982): Irregular growth cycles. *American Economic Review,* **72,** 406-414.

DAY, R.H. (1983): The emergence of chaos from classical economic growth. *Quarterly Journal of Economics,* **98,** 201-213.

DAY, R.H. AND G. PIANIGIANI (1991): Statistical dynamics and economics. *Journal of Economic Behavior and Organization*, **16**, 37-83.

DEBREU, G. (1970): Economies with a finite set of equilibria. *Econometrica,* **38,** 387-392.

DEBREU, G. (1974): Excess demand functions. *Journal of Mathematical Economics,* **1,** 15-21.

DEVANEY, R.L. (1989): *An Introduction to Chaotic Dynamical Sytems*, Redwood City: Addison-Wesley.

DIERKER, E. (1972): Two remarks on the number of equilibria of an economy. *Econometrica*, **40**, 951-953.

DRÈZE, J.H. (1975): Existence of exchange equilibrium under price rigidities. *International Economic Review,* **16,** 301-320.

DROSTE, E.J.R. AND J. TUINSTRA (1998): Evolutionary selection of behavioural rules in a Cournot model: a local bifurcation analysis. CentER Discussion Paper 9886, Tilburg University.

DUFFY, J. (1994): On learning and the nonuniqueness of equilibrium in an overlapping generations model with fiat money. *Journal of Economic Theory,* **64,** 541-553.

EASLEY, D. AND N.M. KIEFER, (1988): Controlling a stochastic process with unknown parameters. *Econometrica*, **56**, 1045-1064.

EVANS, G.W AND S. HONKAPOHJA (1997): Learning dynamics. Forthcoming in *Handbook of Macroeconomics.*

FARMER, R.E. (1993): *The Macroeconomics of Self-Fulfilling Prophecies,* Cambridge: The MIT Press.

FELDMAN, M (1991): On the generic nonconvergence of Bayesian actions and beliefs. *Economic Theory,* **1**, 301-321.

FISHER, F.M. (1961): The stability of the Cournot oligopoly solution: the effects of speeds of adjustment and increasing marginal costs. *Review of Economic Studies*, **28**, 125-135.

FISHER, F.M. (1983): *Disequilibrium Foundations of Equilibrium Economics.* Cambridge: Cambridge University Press.

FRIEDMAN, J.W. (1977): *Oligopoly and the Theory of Games*, Amsterdam: North-Holland.

FRIEDMAN, J.W. (1983): *Oligopoly Theory,* Cambridge: Cambridge University Press.

FURTH, D. (1986): Stability and instability in oligopoly. *Journal of Economic Theory,* **40**, 197-228.

GABSZEWICZ, J.J. AND J.-P. VIAL. (1972): Oligopoly 'a la Cournot' in general equilibrium analysis. *Journal of Economic Theory,* **4**, 381-400.

GARY-BOBO, R.J. (1987): Locally consistent oligopolistic equilibria are Cournot-Walras equilibria. *Economics Letters*, **23**, 217-221.

GALE, D. (1973): Pure exchange equilibrium of dynamic economic models. *Journal of Economic Theory,* **6**, 12-36.

GALE, J., K. BINMORE AND L. SAMUELSON (1995): Learning to be imperfect: the ultimatum game. *Games and Economic Behavior*, **8**, 56-90.

GATES, D.J., J.A. RICKARD AND D.J. WILSON (1977): A convergent adjustment process for firms in competition. *Econometrica,* **45**, 1349-1363.

GATES, D.J., J.A. RICKARD AND D.J. WILSON (1978): Convergence of a market related game strategy. *Journal of Mathematical Economics,* **5**, 79-109.

GAUNERSDORFER, A. AND J. HOFBAUER (1995): Fictitious play, Shapley polygons and the replicator equation. *Games and Economic Behavior,* **11,** 279–303.

GOEREE, J. (1997): *Applications of Discrete Choice Models*, Ph.D. thesis, University of Amsterdam, Tinbergen Institute Research Series 135.

GOEREE, J.K., C. HOMMES AND C. WEDDEPOHL (1997): Stability and complex dynamics in a discrete tâtonnement model. *Journal of*

Economic Behavior and Organization, **33**, 395-410.

GOLUBITSKY, M., I.N. STEWART AND D.G. SCHAEFFER (1988): *Singularities and Groups in Bifurcation Theory*, Vol. II, New York: Springer-Verlag.

GRANDMONT, J.M. (1985): On endogenous competitive business cycles. *Econometrica*, **53**, 995-1045.

GRANDMONT, J.-M. AND G. LAROQUE (1991): Economic dynamics with learning: some instability examples. In: W.A. Barnett et al (eds.) *Equilibrium Theory and Applications*, Cambridge: Cambridge University Press.

GUCKENHEIMER J. AND P. HOLMES (1983): *Nonlinear Oscillations, Dynamical Systems, and Bifurcations of Vector Fields*, New York: Springer-Verlag.

HAHN, F.H. (1962): The stability of the Cournot oligopoly solution. *Review of Economic Studies*, **29**, 329-331.

HAHN, F.H. (1982): Stability. In K. Arrow and M. Intriligator (eds.) *Handbook of Mathematical Economics*, vol II. Amsterdam: North Holland.

HAHN, F.H. AND T. NEGISHI (1962): A theorem on non-tâtonnement stability. *Econometrica*, **30**, 463-469.

HART, O.D. (1985): Imperfect competition in general equilibrium: an overview of recent work. In K. Arrow and S. Honkapohja (eds.) *Frontiers of Economics*, Oxford: Blackwell.

HOMMES, C.H. AND G. SORGER (1999): Consistent expectations equilibria. *Macroeconomic Dynamics*, **2**, 287-321.

JORDAN, J. (1991): Bayesian learning in normal form games. *Games and Economic Behaviour*, **3**, 60-81.

KALAI, E. AND E. LEHRER (1993): Rational learning leads to Nash equilibrium. *Econometrica*, **61**, 1019-1045.

KALAI, E. AND E. LEHRER (1995): Subjective games and equilibria. *Games and Economic Behaviour*, **8**, 123-163.

KEENAN, D. AND T. RADER (1985): Market dynamics and the law of demand. *Econometrica*, **53**, 465-471.

KEHOE, T.J. AND D.K. LEVINE (1985): Comparative statics and perfect foresight in infinite horizon economies. *Econometrica*, **53**, 433-453.

KHIBNIK, A.I., YU.A. KUZNETSOV, V.V. LEVITIN AND E.V. NIKOLAEV (1992): *Interactive LOCal BIFurcation analyzer: LOCBIF*, version 2.

KIEFER, N.M. AND Y. NYARKO (1989): Optimal control of an unknown linear process with learning. *International Economic Review*, **30**, 571-586.

KIRCHKAMP, O. AND K.H. SCHLAG (1994): Endogenous learning rules in social networks. Mimeo, Department of Economics, University of Bonn.

KIRMAN, A.P. (1975): Learning by firms about demand conditions. In: R. Day (ed.), *Adaptive Economics*, New York: Academic Press.

KIRMAN, A.P. (1983): Mistaken beliefs and resultant equilibria. In: R. Frydman and E. Phelps (eds.), *Individual Forecasting and Collective Outcomes*, Cambridge: Cambridge University Press.

KIRMAN, A.P. (1995): Learning in oligopoly: theory, simulation, and experimental evidence. In: A.P. Kirman and M. Salmon (eds.), *Learning and Rationality in Economics*, Oxford: Blackwell.

KOPEL, M. (1997): Simple and complex adjustment dynamics in Cournot duopoly models. *Chaos, solitons, and Fractals,* **7**, 2031-2048.

KUZNETSOV, YU.A. (1995): *Elements of Applied Bifurcation Theory*, Springer-Verlag.

LETTAU, M. AND T. VAN ZANDT (1995): Robustness of adaptive expectations as an equilibrium selection device. Working paper Tilburg University.

LJUNG, L. (1977): Analysis of recursive stochastic algorithms. *IEEE Transactions on Automatic Control,* **22,** 551-575.

LJUNG, L. AND T. SÖDERSTRÖM. (1983): *Theory and Practice of Recursive Identification*, Cambridge, MA: MIT Press.

LUCAS, R.E.,JR. (1971): Econometric testing of the natural rate hypothesis. In: O. Eckstein (ed.), *The Econometrics of Price Determination Conference.* Board of Governors of the Federal Reserve System and Social Science Research Council.

LUCAS, R.J. (1986): Adaptive behaviour and economic theory. *Journal of Business,* **59,** 401-426.

MANSKI, C.F., AND D. MCFADDEN (1981): *Structural Analysis of Discrete Data with Econometric Applications.* Cambridge, MA: MIT Press.

MANTEL, R. (1974): On the characterization of aggregate excess demand. *Journal of Economic Theory,* **7,** 348-353.

MANTEL, R. (1976): Homothetic preferences and community excess demand functions. *Journal of Economic Theory,* **12,** 197-201.

MARCET, A. AND T.J. SARGENT (1989a): Convergence of least squares learning mechanisms in self-referential linear stochastic models.

Journal of Economic Theory, **48**, 337-368.

MARCET, A. AND T.J. SARGENT (1989b): Least squares learning and the dynamics of hyperinflation. In: W.A. Barnett et al. (eds.) *Economic Complexity: Chaos, Sunspots, Bubbles and Nonlinearity.* Cambridge: Cambridge University Press.

MARIMON, R., S.E. SPEAR AND S. SUNDER (1993): Expectationally driven market volatility: an experimental study. *Journal of Economic Theory*, **61**, 64-103.

MARSCHAK, T. AND R. SELTEN (1974): *General Equilibrium with Price-Setting Firms.* New York: Springer-Verlag.

MCMANUS, M. AND R.E. QUANDT (1961): Comments on the stability of the Cournot oligopoly solution. *Review of Economic Studies*, **28**, 136-139.

MORA, L. AND VIANA, M. (1993): Abundance of strange attractors. *Acta Mathematica*, **171**, 1-71.

MUTH, J.F. (1961): Rational expectations and the theory of price movements. *Econometrica*, **29**, 315-335.

NEGISHI, T. (1961): Monopolistic competition and general equilibrium. *Review of Economic Studies*, **27**, 136-139.

NEWHOUSE, S. (1974): Diffeomorphisms with infinitely many sinks. *Topology*, **13**, 9-18.

NEWHOUSE, S. (1979): The abundance of wild hyperbolic sets and nonsmooth stable sets for diffeomorphisms. *Publ. Math. I.H.E.S.*, **50**, 101-151.

NIKAIDO, H. (1975): *Monopolistic Competition and Effective Demand.* Princeton University Press.

NOVSHEK, W. (1985): On the existence of Cournot equilibrium. *Review of Economic Studies*, **52**, 85-98.

NOWAIHI, A. AL- AND P.L. LEVINE (1985): The stability of the Cournot oligopoly model: a reassesment. *Journal of Economic Theory*, **35**, 307–321.

NYARKO, Y., (1991): Learning in mis-specified models and the possibility of cycles. *Journal of Economic Theory*, **55**, 416-427.

NYARKO, Y., (1994): Bayesian learning leads to correlated equilibria in normal form games. *Economic Theory*, **4**, 821-841.

PALIS, J. AND F. TAKENS (1993): *Hyperbolicity and Sensitive Chaotic Dynamics at Homoclinic Bifurcations.* Cambridge: Cambridge University Press.

POINCARÉ, H. (1890): Sur le problème des trois crops et les equations de la dynamique. *Acta Mathematica.* **171**, 1-171.

RAND, D. (1978): Exotic phenomena in games and duopoly models. *Journal of Mathematical Economics*, **5**, 173-184.

REICHLIN, P. (1986): Equilibrium cycles in an overlapping generations economy with production. *Journal of Economic Theory*, **40**, 89-102.

RHODE, P. AND M. STEGEMAN (1995): Non-Nash equilibria of Darwinian dynamics. Mimeo, Department of Economics, University of North Carolina.

ROBERTS, D.J. AND H. SONNENSCHEIN (1977): On the foundations of the theory of monopolistic competition. *Econometrica*, **45**, 101-114.

SAARI, D.G. (1985): Iterative price mechanisms. *Econometrica*, **53**, 1171-1131.

SAARI, D.G. AND C.P. SIMON (1978): Effective price mechanisms. *Econometrica*, **46**, 1097-1125.

SAMUELSON, P.A. (1958): An exact consumption-loan model of interest with or without the social contrivance of money. *Journal of Political Economy*, **66**, 467-482.

SAMUELSON, L. (1997): *Evolutionary Games and Equilibrium Selection.* Cambridge, MA: MIT Press.

SARGENT, T.J. (1993): *Bounded Rationality in Macroeconomics*, Oxford: Clarendon Press.

SARGENT, T.J. (1998): *The Conquest of American Inflation*, Princeton: Princeton University Press.

SCARF, H. (1960): Some examples of global instability of the competitive equilibrium. *International Economic Review*, **1**, 157-172.

SCHINKEL, M.P., J. TUINSTRA AND D. VERMEULEN (1999): Convergence of Bayesian learning in mis-specified models. In preparation.

SCHLAG, K.H. (1996): Why imitate, and if so, how? a bounded rational approach to multi-armed bandits. *Journal of Economic Theory*, **78**, 130-156.

SCHÖNHOFER, M. (2000): Chaotic learning equilibria. *Journal of Economic Theory*, **89**, 1-20.

SEADE, J. (1980): The stability of Cournot revisited. *Journal of Economic Theory*, **23**, 15-27.

SHAPLEY, L. (1964): Some topics in two-person games. *Annals of Mathematical Studies*, **4**, 1-28.

SILVESTRE, J. (1977): A model of general equilibrium with monopolistic behavior. *Journal of Economic Theory*, **16**, 425-442.

SMALE, S. (1976): A convergent process of price adjustment and global Newton methods. *Journal of Mathematical Economics*, **3**, 107–120.

SONNENSCHEIN, H. (1973): Do Walras' identity and continuity characterize the class of community excess demand functions? *Journal of Economic Theory*, **6**, 345-354.

TAKAYAMA, A. (1985): *Mathematical Economics* (2nd edition). Cambridge: Cambridge University Press.

THALER, R. (1992): *The Winner's Curse: Paradoxes and Anomalies of Economic Life.* New York: Free Press.

THEOCHARIS, R.D. (1959): On the stability of the Cournot solution on the oligopoly problem. *Review of Economic Studies*, **27**, 133-134.

TIROLE, J. (1988): *The Theory of Industrial Organization.* Cambridge, MA: MIT Press.

TUINSTRA, J. (1997): A price adjustment process with symmetry. *Economics Letters*, **57**, 297-303.

TUINSTRA, J. (2000): A discrete and symmetric price adjustment process on the simplex. *Journal of Economic Dynamics and Control*, **24**, 881-907.

TUINSTRA, J. AND C. WEDDEPOHL (1999): On the equivalence between overlapping generations models and symmetric general equilibrium models. *Journal of Economics*, **70**, 187-207.

UZAWA, H. (1959): Walras' tâtonnement in the theory of exchange. *Review of Economic Studies*, **27**, 182-194.

UZAWA, H. (1962): On the stability of Edgeworth's barter process. *International Economic Review*, **3**, 218-232.

VEGA-REDONDO, F. (1996): *Evolution, Games, and Economic behavior.* Oxford: Oxford University Press.

VEGA-REDONDO, F. (1997): The evolution of Walrasian behavior. *Econometrica*, **65**, 375-384.

DE VILDER, R. (1995): *Endogenous Business Cycles.* Thesis, Department op Economics, Tinbergen Institute Research Series 96, University of Amsterdam.

DE VILDER, R. (1996): Complicated endogenous business cycles under gross substitutability. *Journal of Economic Theory*, **71**, 416-442.

WEDDEPOHL, C. (1995): A cautious price adjustment process on the simplex. *Journal of Economic Behavior and Organization*, **27**, 293-300.

WEIBULL, J.W. (1995): *Evolutionary Game Theory.* Cambridge, MA: MIT Press.

WOODFORD, M. (1990): Learning to believe in sunspots. *Econometrica*, **58**, 277-307.

YORKE, J.A. AND ALLIGOOD, K.T. (1983): Cascades of period doubling bifurcations: a prerequisite for horseshoes. *Bulletin American Mathematical Society,* **9,** 319-322.

YOUNG, H.P., AND FOSTER, D. (1991): Cooperation in the short and in the long run. *Games and Economic Behavior,* **3**, 145-156.

ZELLNER (1971): *An Introduction to Bayesian Inference in Econometrics.* New York: John Wiley & Sons.

Index

GPSR Compliance

The European Union's (EU) General Product Safety Regulation (GPSR) is a set of rules that requires consumer products to be safe and our obligations to ensure this.

If you have any concerns about our products, you can contact us on ProductSafety@springernature.com

In case Publisher is established outside the EU, the EU authorized representative is:

Springer Nature Customer Service Center GmbH
Europaplatz 3
69115 Heidelberg, Germany

Batch number: 10837190

Printed by Printforce, the Netherlands